The Policy Process

The Policy Process
A Practical Guide for Natural Resource Professionals

Tim W. Clark

Yale University Press

New Haven and London

Published with assistance from the foundation established in memory of Philip Hamilton McMillan of the Class of 1894, Yale College.

Set in Aster Roman type by Achorn Graphic Services, Worcester, Massachusetts.
Printed in the United States of America.

Library of Congress Cataloging-in-Publication Data

Clark, Tim W.
    The policy process : a practical guide for natural resource professionals / Tim W. Clark.
        p.   cm.
    Includes bibliographical references and index.
    ISBN 0-300-09011-0 (cloth : alk. paper) — ISBN 0-300-09012-9 (pbk.)
    1. Conservation of natural resources—Decision making.   I. Title.
S944.5.D42 C63 2002
333.7'2—dc21

                                                                    2001046637

A catalogue record for this book is available from the British Library.

The paper in this book meets the guidelines for permanence and durability of the Committee on Production Guidelines for Book Longevity of the Council on Library Resources.

10 9 8 7 6 5 4 3 2 1

The significant problems we face cannot be solved at the same level of thinking we were at when we created them.

Albert Einstein

The way we *see* the problem *is* the problem. . . . We need to understand our own "paradigms" and how to make a "paradigm shift.". . . We need a new level, a deeper level of thinking—a paradigm based on the principles that accurately describe the territory of effective human being and interacting—to solve these deep concerns.

Stephen R. Covey

We think in generalities, we live in detail.

Alfred North Whitehead

If you are thinking a year ahead, sow seeds. If you are thinking ten years ahead, plant a tree. If you are thinking a hundred years ahead, educate people.

Kuan-Tsu, 3d century B.C.

Contents

Preface

All natural resource professionals participate in policy processes in various ways, at various scales, and with various degrees of success, whether they are aware of it or not. This book shows professionals how they can improve their understanding of and effectiveness in this process. It is meant to complement existing courses and university texts in wildlife biology, conservation biology, forestry, range management, ecosystem management, sustainable development, and related subjects, or it can stand alone in a separate course. It can also be used in seminars, in workshops, and in analysis of actual cases involving a broad range of issues. Practicing professionals in many fields will find it useful to gain further perspective on their work. The aim of the book is to improve both the substance and the process of natural resource conservation, whether it is focused on single species, whole ecosystems, sustainability projects, government programs, nongovernmental initiatives, international development, or something else. Knowledgeable, policy-oriented leadership is needed in this important work, and it must be informed by a solid, functional understanding of the policy process.

This book describes the conservation process using the oldest and most comprehensive theory and method available in the modern policy analytic movement—the policy sciences. The policy sciences are simultaneously a theory about society and a method of inquiry into problems and associated social and decision processes. The book introduces practical, analytic concepts and skills for improving the outcomes of these processes. This approach, which helps professionals understand

and deal with the rationality, politics, and morality of any problem they confront and promotes a functional understanding of the policy process applicable to all contexts, supports the overriding goal of democracy or a commonwealth of human dignity. Although learning the method and becoming skilled in its use require experience and reflective practice, even a little knowledge of the policy sciences and how to apply them can result in significant improvements in resource management. The case material uses efforts to conserve biological diversity to illustrate the policy process since this is such an important theme globally.

Two audiences are targeted. The first consists of university students beginning their professional development who need to be equipped with the best tools to enter the workforce, whether they set their sights on work in government or nongovernment sectors. The second consists of on-the-job professionals who want to improve their understanding of the policy process and their effectiveness in it. These people are already active policy participants who desire a fresh vision to see what it all means and how to organize their considerable experience for improved performance.

The policy sciences, a well-respected area of knowledge more than fifty years old, had part of its origin in John Dewey's philosophy of "American Pragmatism" and his 1910 book *How We Think*. The initial effort to devise the conceptual framework of the policy sciences was led by Harold Lasswell in the decade just before World War II. It was refined, streamlined, and abstracted throughout the 1950s and 1960s. One scholar called Lasswell a "one-man university" whose "competence in, and contributions to, anthropology, communications, economics, law, philosophy, psychology, psychiatry, and sociology are enough to make him a political scientist in the model of classical Greece" (cited in Lasswell and McDougal 1992, xxxii). He is among the most creative and productive social scientists of the twentieth century. In 1971 Lasswell published *A Pre-view of the Policy Sciences*, a brief but relatively comprehensive articulation of the richness and potential of this policy approach. Among other central writings are Lasswell and Abraham Kaplan, *Power and Society: A Framework for Political Inquiry* (1950) and Lasswell and Myres McDougal, *Jurisprudence for a Free Society: Studies in Law, Science, and Policy* (1992). A large body of literature elaborates the scope and applications of the policy sciences (see Clark et al. 2000a).

My own understanding of the policy sciences clearly reflects my history, which has included formal studies in zoology and related subjects, field work on a number of vertebrate species and scientific and management issues, involvement in endangered species and ecosystem management programs, consulting on international conservation issues, and professional education. Teaching and working in the biological conservation field have brought me into contact with students, professionals, and officials from more than thirty countries. My graduate students at the Yale School of Forestry and Environmental Studies have come from varied backgrounds—conservation biology, forest management, natural resource economics, engineering, environmental law, public affairs, organization and management, international relations, anthropology, sociology, planning, and environmental policy with various emphases on marine issues, freshwater and wetlands, wildlife and biodiversity, sustainable development, or other subjects. Many of the people with whom I have worked, including students, have had years of on-the-job experience in government agencies or in nongovernmental organizations. Their professional work has demanded that they become increasingly adept in solving problems. They have needed to be explicit, comprehensive, and practical. And nearly all the people who have been exposed to the interdisciplinary approach of the policy sciences and have delved into its concepts and categories have found it to be an extremely helpful guide to solving complex natural resource problems.

I have a number of debts that I gladly acknowledge. Intellectually, I owe a great deal to Garry D. Brewer, Ronald D. Brunner, Andrew R. Willard, and William Ascher, with whom I have had numerous discussions over the years. These four were either students or professional colleagues of Lasswell and McDougal. Many other people also helped me to better understand this approach and its practical value.

I owe a professional debt to the many people who provided me with opportunities to work on diverse management and policy issues and contexts in the classroom and in the field in the United States, Canada, Australia, Indonesia, China, Inner Mongolia, Belize, Panama, Costa Rica, and other countries. Important in this regard are Robert J. Begg, Mark Ashton, Gary N. Backhouse, Kim W. Lowe, Nicki Mazur, Steve Cork, Peter Stroud, Barbara Wilson, and Peter Myroniuk.

In terms of education, I am indebted to several valuable colleagues, especially at Yale University, including many masters and doctoral students in my courses and seminars. My colleagues have included Stephen Kellert, John Wargo, William Burch, Leonard Doob, F. Herbert Borman, Richard Burroughs, Michael Dove, and John Gordon. Colleagues outside the Yale community include David Mattson and Steven Primm. Doctoral students with whom I have interacted and who have gone on to use the policy sciences in recent years include Richard Reading, Richard Wallace, Murray Rutherford, Christina Cromley, Peter Wilshusen, Eva Garen, Alejandro Flores, and Matthew Auer.

Generous funding has come from Deans John Gordon and Jerry Cohon at the Yale School of Forestry and Environmental Studies and Director Don Green at Yale's Institution for Social and Policy Studies. Many other people have supported my work through my affiliation with the Northern Rockies Conservation Cooperative in Jackson, Wyoming. Deserving of special thanks are Peyton Curlee Griffin (former executive director), Cathy Patrick, Hopie and Bob Stevens (Fanwood Foundation), Gil Ordway, Sybil and Tom Wiancko (Wiancko Charitable Foundation), Kathe Henry (Scott Opler Foundation), Ann Harvey (New-Land Foundation), and Ted Smith and Gary Tabor (Kendall Foundation).

Also significant is the debt I owe my wife, Denise Casey, whose tireless editorial hand is everywhere evident in this book. Finally, several people critically reviewed all or parts of the book, and I am most grateful for their time and comments. They are Andrew Willard, Murray Rutherford, Christina Cromley, Nicki Mazur, Robert Hoskins, William Ascher, Robert Begg, Kim Lowe, and three anonymous reviewers chosen by Yale University Press.

To all these people, thank you.

1 Introduction
Professional Challenges

The work of professionals is to apply their special knowledge and skills responsibly in resolving societal problems in the common interest. This is not a simple, straightforward task. As George Albee (1982) noted, the more professionals become involved in real-world problems, the more politically enlightened they become, realizing that actual prevention of environmental misuses must involve social and political changes. These dimensions of management and policy are by far the most complex and challenging part of being a natural resource professional.

This chapter characterizes these challenges and the limitations of conventional problem solving and introduces the policy process within which management and conservation of natural resources take place. It then describes a *policy orientation* for professionals that will allow them to address nonscientific variables along with scientific ones and thus, with a more comprehensive understanding of the problems and solutions, better serve society's common interests.

The Need for New Thinking

It is easy to become overwhelmed or discouraged by multiple demands, rapid changes, decreasing financial resources, and the complexity of natural resource management issues. In the United States, for example, there are proposals for agency reorganizations, downsizing, and new initiatives, including shifts to ecosystem management, by nearly all the federal agencies. These are accompanied by range, wildlife, and forest management reforms on public lands and new working relationships with the public and business. All these changes put new demands on professional practice. At such times the evolving mission of professionals may be unclear, leadership muddled or lacking, day-to-day activities unsupported or unrewarded, and practical ways to achieve goals frustrated or absent. On a personal level, working on issues that involve significant conflict can be draining. Those representing special interests push and pull for different outcomes or policies; values, visions, and means are at odds in many instances. Government administrations come and go at state (or provincial) and national levels, sometimes causing dramatic turnabouts in policy, programs, and on-the-ground actions. These, too, affect professionals for better or worse.

What are professionals to make of this changing world of practice? How can they best understand these processes in order to be most effective? What are they to think of the conflict surrounding forest management, about the rhetoric over endangered species policy, or about the controversies attending pollution abatement? How do practitioners negotiate these dynamic social processes and chart a reasonable, justifiable personal and professional direction for themselves? How can they contribute to organizations that seek advancement, improvement, sustainability, and society's common interests? Much of the complexity of natural resource policy appears to be beyond comprehension or remediation.

The usual techniques, invoked automatically when problematic situations arise, often prove unsatisfactory and may even confound matters. Alan Miller (1999) argues strongly that conventional problem solving is itself the problem and that reform of its inadequacies is desperately needed. Conventional approaches tend to simplify policy problems, misconstrue some vital part of the context, or overlook the context altogether (Brunner 1991a, 1996a; Clark 1993). And professionals come to realize—often in hindsight, after action has been taken, commitments made, and results have disappointed—that inadequate problem solving resulted in suboptimal outcomes. One example of this was the Greater Yellowstone Coordinating Committee's "vision" exercise for the Yellowstone region in the late 1980s and early 1990s (Lichtman and Clark 1994). With a mandate to integrate agency actions with a common management vision, the committee apparently thought it was on the right track in preparing a general plan. After committing substantial resources, members found that their assumptions were wrong when both conservationists and traditional commodity extraction industries overwhelmingly rejected their document. The agencies went back to the drawing board and eventually published a watered-down version of little practical use. In another high-profile case in the Yellowstone area, bison management remained unresolved after fifteen years, the expenditure of millions of dollars, and numerous lawsuits (Cromley 2001). In both these cases opportunities to find and serve common interests were lost because of the inabilities of participants to analyze the policy problems fully and interpret the contexts accurately.

This tendency to simplify, misconstrue, or overlook occurs because professionals (and their organizations) may be unduly preoccupied with or entrapped by some mental construct such as experimental science, positivism, disciplinary boundaries, bureaucratic procedures, job descriptions, program boundaries, or policy preferences (Simon 1983, 1985; Brunner 1995a). If they are overly trained or predisposed to these bounded patterns or models, they may not pay adequate or realistic attention to the context of their activities. This problem is widespread and deeply rooted. Too often problems in the management and policy of natural resources are viewed only (or largely) within very narrow conceptions of "positivistic" science, for instance, thus creating the misperception that only the biological sciences are required to solve them (Brewer and Clark 1994).

According to the traditional view, science is understood to be value-free, objective, and reductionistic in its search for universal laws, its use of formal rigor, and its application of quantitative precision. This kind of science is, without a doubt, good at solving certain kinds of problems, especially those that can be controlled, and its success is largely the result of the great care scientists take in selecting problems. But toxic wastes, ecosystem degradation, and species endangerment, for instance, are not problems selected by scientists and investigated under carefully controlled conditions. Nevertheless, scientists (and managers using science) are charged with fixing them (and are often blamed for their existence). A great many of the conservation and management problems that exist today fall largely outside the bounds that conventional science has established for itself. In these cases, the special kind of rationality, discipline, and quantitative methods devised to support the conventional scientific edifice provide little help (Brewer and deLeon 1983). The philosophy and methods of experimental science are not sufficient—even though they may be necessary—for their resolution (Funtowicz and Ravetz 1990; Brunner and Clark 1997). But too often we invoke this kind of science without deliberating about whether it is appropriate.

This assertion should not be misconstrued as antiscientific. Experimental (positivistic) science must and will continue to help. But it cannot supply many of the essential elements of successful problem solving in public policy. We must recognize when and how conventional science can and cannot help. Natural resource problems are addressed only out of necessity; scientists, managers, administrators, policy makers, and citizens take on these problems as they are thrust upon them. In these kinds of public policy situations, another kind of science and problem solving is clearly needed, and natural resource professionals need different tools to meet these challenges.

Policy-Oriented Professional Practice

The policy sciences offer professionals a practical guide to dealing with real-life events in all their complexity. Natural resource problems involve social complexity, people's perceptions and values, and often significant amounts of uncertainty. In these kinds of problems no objective, scientifically verifiable, optimal solution is possible, yet the problems are real and demand solution. Professionals face problems that cannot be controlled and carefully selected, in some ways putting them at a disadvantage compared to field biologists, for example, who have a traditional (albeit limited) tool kit of experimental science at their disposal.

The *policy sciences* consist of a set of integrated concepts or conceptual tools for framing thought and action and for guiding analysis, interpretation, and resolution of any problem (Lasswell 1968). The *sciences* part of the term refers to systematic, empirical inquiry. The concepts focus our attention on three key questions that can be asked about any policy, proposal, or initiative (Brunner, personal communication): Is it rational? Is it politically practical? Is it morally justified? This framework suggests additional important questions: From whose standpoint is the policy problem best understood? What methods are required to understand the problem? How should answers to these questions be integrated with ongoing practices? These interrelated questions should be asked and addressed in every management and policy case.

The basic policy sciences concepts are useful, according to Ronald Brunner (1995a), in finding the important pieces of information in a maze of reports that are typically distorted and incomplete, identifying what important pieces are missing, organizing the pieces into a coherent picture, evaluating that picture from rational, political, and moral standpoints, and devising a policy of your own or constructing a more educated opinion of the policy under question. You will learn from using the policy sciences that many natural resource programs and policies are deficient in important ways and that you can contribute directly to improving them.

Mastery of the basic policy sciences concepts requires study and practice, and, very likely, unlearning some of what you learned from past training and experience. But their continued use in many different kinds of problematic situations can produce a knowledgeable, skilled, and successful professional. The practical pay-off will be understanding, insight, and successful outcomes for the sustainability of natural resources, the benefit of society, and the enrichment of your professional practice. These ideas can help you step outside your conventional frames of reference and develop a new perspective on yourself and the social processes in which you are engaged. They can help increase the visibility of relevant events concerning a problem and its context and thus help you and your co-workers become better problem solvers and meet your civic responsibility to solve society's environmental problems in a manner that reflects the common interest.

The Policy Process

The policy process is a social dynamic that determines how the good and bad in life are meted out—that is, who gets what, when, and how (Lasswell 1950a). Like philosophy, policy wrestles with fundamental problems about people, how we live, how we find meaning, and how we go about making important decisions (see Parsons 1995). Policy is a process focused on problem-solving, usually involving some technical content (the substance, such as regional assessments, reauthorization debates for the Endangered Species Act, or ozone depletion) and always involving people with varying perspectives and interests in the problem and its solution (the process). William Ascher and Robert Healy (1990, 159–60) offer a particularly useful definition:

> Policymaking is a sequence of many actions by many actors, each with potentially different interests, information, roles, and perspectives. . . . No one can guarantee that policy will 'optimize for the system as a whole'—although there may be institutions (such as planning agencies) created to do so. Each of the phases of the policy process is populated by somewhat different official agencies and interest groups; each faces different analytic and political challenges. . . . Their efforts to coordinate or to contest with one another, and their limitations in grappling with the complexity of policy dilemmas, are typically left out of the texts on natural resource economics and management. Yet these interactions, what we call the 'policy process,' often spell the difference between success and failure in resource use.

There are many misconceptions about the term *policy* in natural resource fields. It is sometimes thought to be synonymous with politics; terms such as biopolitics embody this view. It is also common to hear resource professionals talk about science versus politics, with the implication that politics is bad and science is good and that if we had more science and less politics life would somehow be better. Another misconception is to equate policy with a plan, mission, goal, or law. Hogwood and Gunn (1986, 13–19) distinguish ten ways in which the term policy is commonly used, all of which can be observed in any newspaper over the course of a few weeks: (1) a field of study, such as wildlife policy, (2) an expression of general purpose or desired state of affairs, as in "we shall endeavor to restore endangered species," (3) a specific proposal, such as "we shall establish ten populations," (4) a decision of government, including specific, on-the-ground management decisions, (5) formal authorization, such as the Endangered Species Act, (6) a program, as in "our policy is to set up public-private partnerships," (7) output, or what government delivers, (8) outcome, or what is actually achieved, (9) a theory or model, such as "assumptions about cause and effect relationships" about a problem and how it should be solved, and (10) a process, as of complexities unfolding over time.

Care should be taken in using the term. *Policy,* as used in this book and following Lasswell and McDougal (1992), is a social process of authoritative decision making by which the members of a community clarify and secure their common interests. In other words, the people who interact in a community share expectations about who has the authority to make decisions about what, when, and how. According to Brunner (1996a, 46), ultimate authority in society to make policy rests "in the perspectives of living members of the community—their identifications, demands, and expectations—which, like other factors in social process, are amenable to empirical inquiry." The policy sciences can help professionals conduct this vital "empirical inquiry" into people's perspectives, interactions, and the outcomes of decision-making processes.

Natural Resources

The ongoing interaction of people in their efforts to achieve what they value is the *policy process.* It is the never-ending, value-laden efforts of people to organize themselves effectively to solve important collective problems and find meaning for themselves. One such problem is the environment and the use of natural resources. Even though all people value a healthy environment, different people and groups in society want to use resources in different ways, under the conviction that different practices with different results will best serve the common (or their own) interests. Making decisions about natural resources is thus part of larger societal processes and reflects the same patterns of human behavior manifest in other policy arenas. The two fundamental questions about natural resource policy and management are: How are we going to use natural resources? And who gets to decide? A community's struggle to work out answers to these questions is the making of policy. The role of the professional is to help address these questions in ways that effectively serve the community's interests—that is, in ways that are rational, politically practicable, and justifiable. The policy sciences offer a theory and a conceptual framework that can help professionals understand the complexity, the confusion, and the conflict that often characterize the search for answers to these two questions.

Resources is the term used to designate the physical environment in which human interactions are carried on and which may be directly involved in such interactions (Lasswell and McDougal 1992; Clark, Willard, and Cromley 2000). In the broadest sense, natural resources include the earth and its atmosphere, soils, minerals, energy, plants, animals, and humans as well as the solar system, the galaxy, and beyond. The significance or meaning of natural resources is influenced by the values people seek at various times and places and by their expectations about how they can use resources to achieve their values. For example, a forest is a resource that can be used for well-being (solitude, feeling good, building homes), enlightenment (knowledge about ecosystem structure), wealth (timber sales), or other values.

Humans manipulate their environments, and such *operations* are interwoven with the entire social and cultural context. Manipulating flint for stone tools, atoms for hydrogen bombs, or soils, plants, animals, or ecological processes (such as fire) for agriculture are all examples of operations. When operations are associated with meaning or value (that is, people's perspectives), which is usually the case, we call them *practices*. For example, mining gold may be associated with minimizing pollution (that is, skill), making money (that is, wealth), or some other mix of values. Different cultures carry out operations differently. Some operations are intended for short-term value accumulation, whereas others are long-term; we often call the former *exploitive* and the latter *sustainable* practices. *Technology* is the means people use to manipulate their cultural materials. The objects of manipulation in natural resource management are either raw materials or cultural materials. Raw materials are potentially usable resources, such as un-exploited grasslands, species, oceans, or even the Antarctic. Cultural materials, which change over time, include axes, plows, bulldozers, domestic livestock and crops, bureaucracies, satellites, and computer software—even some soils and whole landscapes if they have been altered significantly by human cultivation. Thus the material culture of a society or community includes its technology and the physical objects or cultural materials on which technology operates.

A Practical Theory of Problem Solving

We know that many "nonscientific" variables are at play in the policy, management, and conservation of natural resources, such as changes in the ways humans value nature and resources, the dynamics of interest groups, and agency organizational cultures. Other variables that are hard to account for—yet vital—include the epistemological (knowledge) and cognitive (thinking) perspectives of professionals, standards for successful problem solving, and conceptions about the proper relation between professionals and society. These matters are intrinsic to natural resource problems, and even though they are not easily measured or controlled, as Ron Westrum (1994) quipped, these soft variables have hard consequences. Good practitioners need to take these variables into account in addition to the more technical ones that they typically consider.

Society spends tremendous resources trying to find solutions to problems of global warming, biodiversity loss, landscape fragmentation, and innumerable others, large and small. These kinds of problems, with important political, economic, and social dimensions, can be immensely intricate and resistant to solutions based on traditional frames of reference. Government agencies do not appear adept at solving such problems quickly or flexibly, academic theorists cannot provide the answers, and idealistic environmentalists are at a loss as well. "Despite the considerable temptations of the optimal or best solutions the conventional disciplines seem able to provide, the information they treat is selective and treats only the what-has-been, not the what-will-be that most concerns policymaking and -makers," say Brewer and de-Leon (1983, 4–5).

Whatever the solutions to resource problems, they must reflect not only technical but also social, political, and institutional considerations, all integrated into a composite picture of the whole problem and its context. All these considerations reflect the values that people attribute to nature, a complex subject that science has partially illuminated. The critical task for professionals, therefore, rather than scientific analysis or prediction, is to incorporate in decision making all the relevant human perspectives and values to the extent consistent with open democracy. Any theory or guideline for action must recognize the centrality of human values in all aspects of managing and conserving nature, and it must offer professionals a practical means to understand values and participate with others in developing sound policy. Science alone offers no methods for integrating the broad range of relevant data into context-specific information of the type required for intelligent management and policy decisions.

An Analytic Framework

Professionals need a practical theory that can be used to model, or "map," any policy process accurately. Just as we need maps to locate ourselves spatially in an ecosystem and to carry out management activities, we need realistic maps of the policy processes in which we participate. Professional schooling teaches sophisticated skills in reading and making landscape maps and in layering and integrating data. It also provides necessary concepts and language to talk about maps and mapmaking. Similarly, there are skills, concepts, and language we can learn to construct useful maps of policy processes (Brunner 1997a, b).

The Framework

The policy sciences conception of social and policy processes is abstracted in a framework consisting of a logically complete set of mapping categories that can help us understand and resolve any policy problem. It is a practical means of organizing our thinking, our knowledge, and our problem-solving efforts, thus allowing us to define a problem and understand its context. Like any good cartographic tool, the framework helps us map a route through a "policyscape" to a destination, highlighting key features and revealing both pitfalls and opportunities.

The framework groups variables into three principal dimensions (figure 1.1), which are elaborated in the following chapters: (1) *social process,* which is mapped in terms of participants, perspectives, situations, base values, strategies, outcomes, and effects; values (of which there are eight—power, wealth, enlightenment, skill, rectitude, respect, well-being, and affection) are the key elements in understanding people's behavior and interactions; (2) *decision process,* which is mapped in terms of seven functions—intelligence, promotion, prescription, in-

Fig. 1.1.   The principle dimensions, categories, and terms of the policy sciences approach to problem solving organized into a framework

vocation, application, appraisal, and termination; and (3) *problem orientation,* which comprises the analytical tasks of clarifying goals, describing trends, analyzing conditions, projecting developments, and inventing, evaluating, and selecting alternatives.

There are three additional key features of the policy sciences approach, each, again, addressed in a separate chapter. First, the *standpoint* of the participant or observer doing the analysis needs to be established in relation to the policy process. Second, *multiple methods* must be used to gather, interpret, and integrate information as a basis for making and carrying out decisions. And third, a single moral goal should guide professionals' problem-solving efforts—what Lasswell and McDougal (1992) call "a commonwealth of human dignity," or, in more popular terms, democracy, human rights, or security. These six elements are the policy sciences framework in a nutshell.

The framework is primarily an effort to systematize the major variables with which social scientists grapple in all political and policy inquiry. The founders of the policy sciences wanted to devise a theory whose interrelated concepts and categories would be "dependable, appropriately selective, creative and economic," (Lasswell and McDougal 1992, 4; Brunner 1996a). The framework does just that, serving as both a theory and a procedure for decision making, for inquiry into the policy process, for orienting ourselves to problems in context, and for understanding our own roles and standpoints in all these situations (Moore 1968). It thus functions as a "stable frame of reference" that can be applied to any case since it does not impose any conventions or ideologies. It simply allows us to focus on and integrate whatever is appropriate in the actual context and stimulate creative insight into how we can bring about more rational action.

The framework also outlines principles of critical thinking. Because it is comprehensive yet sufficiently precise to guide attention to all the significant features of the policy process, it helps us avoid the tendency to depend on fragments of knowledge, single disciplinary views, or ideological stands to define policy problems (Brunner 1996a). It helps us avoid the problems of the three blind men in the well-known tale who each tried to define an elephant by touching only part of its body; one man felt the trunk ("an elephant is like a snake"), another a leg ("it is like a tree"), and the third its side ("it is like a wall"). These "definitions," lacking detail, realism, wholeness, and utility, warn us of the dangers of selective observation and knowledge. The mapping categories of the framework draw attention to all aspects of a policy dynamic, not only selected parts. It is systematic rather than eclectic. It is reliable and operational. Any of the mapping categories in the framework can be a starting point for empirical inquiry, but all of them should be thoroughly investigated.

Benefits of the Framework

One of the primary benefits of adopting this stable frame of reference is that it can give users a functional understanding of policy processes. Those of us trained in biology, physiology, or ecology already appreciate the significance of understanding systems functionally. For example, we employ an evolutionary or functional classification scheme to group plants and animals in order to understand their interrelations and selective outcomes over long periods. We also see a landscape in ecological, or functional, terms—dynamic vegetative communities, land use changes through time, stream flows, succession, predation, biogeochemical cycling, or energy flows—in contrast to laymen, who see it only in conventional terms of attractiveness, resources, utility, ownership, undifferentiated patterns of forests, grasslands, or bodies of water. They do not see it as a system in flux and do not know how it functions as a whole, how its parts are interrelated, or what forces drive it.

Many professionals trained in natural resource fields have a similarly limited, conventional view of policy making. We often do not understand how it functions as a system, what consequences it has in people's lives, or what practical influence we might have on it. We take things at face value and cannot think critically about the interrelations of the components or the driving forces. As a result, it seems disturbingly meaningless and capricious. Without useful concepts or language with which to talk about the policy process, we dismiss people and events with stock phrases such as "political," "conflict-ridden," or "unethical," and we explain the failure of the system to achieve sustainable resource management (or any goal) by resorting to the same old causes such as lack of money, outside interference, or too little research. According to the historian Richard Slotkin (1992), conventional perspectives do not allow us to engage difficult problems fully. In fact, they replace the troublesome and problematic facts of the real world with a counterworld of pseudofacts, images, symbols, and formulae.

The policy sciences framework and its associated theory, however, foster a functional perspective that allows us to overcome limited, conventional problem-solving approaches. Just as ecological knowledge allows resource managers to carry out their activities with sophistication, so does the framework allow its users to understand and participate in policy processes rationally, practically, and justifiably. The practical advantage of a functional outlook is that it maximizes the likelihood that participants will have influence and power over policy outcomes, and, for natural resource professionals, this means more successful management and more rewarding careers.

The mapping categories of the policy sciences framework are constructed to meet five "design criteria" that can help you appreciate their purpose and use—and help you evaluate alternative frameworks (Lasswell and McDougal 1992; Brunner 1996a). First, the categories are both selective and comprehensive, allowing you to orient yourself to the problem and context at hand. You want to focus on a particular problem yet keep the big picture in mind at the same time. Second, the list of categories is short, concise, and simple, making it easy to use. Third, the categories are appropriately flexible, permitting many more refined distinctions than do conventional approaches. Like the step-down menus of computer programs, the categories give you access to successively more detailed and specific "menus" as you need them. Fourth, the categories are used heuristically to guide your attention to the pertinent features of the relevant contexts. Instead of showing exact equivalencies between the categories and the specific case under investigation, they help you discover, be creative, and open up your inquiry into any problem. They are adequate for describing the dynamism of policy processes, yet they provide a structure for analysis and decision making. Fifth, the categories are highly abstract in order to facilitate broad usage. The concepts do not refer to any particular situation. Their principal purpose is to help you set aside preconceptions, including your ideology, about the problems you in which are interested.

There are certain things the policy sciences do not do. They provide no quick fixes to the numerous complex policy problems that exist—in natural resources or in any other field of human interaction. Nor are they a neat technique for marrying analytic methods to democratic or participatory public processes. And, although the policy sciences recommend or offer a "correct way" to make decisions with procedural standards, they do not provide a clear indication of the correct, best, or optimal substantive content of decisions (Lasswell and McDougal 1992). The framework provides questions worth asking and an agenda for asking them. The answers will depend on the context and on the participants' ability to make connections between the concrete situation being mapped and the abstract, "open" categories of analysis. The policy sciences framework also does not allow you to be definitive or final because the context—what you are trying to understand and map—is in constant change (partly in response to your actions to improve it). There are always more decisions to be made and more needed, yet, at some point, participants must decide that they have acquired enough data and analysis and spent enough time and resources, given the circumstances, for their purposes. Finally, although the policy sciences provide useful language, the exact terms people use are not as important as the concepts themselves and people's translations of them, empirical references to real situations, and their self-consistency (Lasswell and Kaplan 1950).

The Concept of the Common Interest

Interests of all kinds are at the heart of natural resource policy and management. The concepts of common and special interests are well known to most people and have been described and illustrated by many scholars (McDougal et al. 1981; Lasswell and McDougal 1992; Shutkin 1999; Warren 1999). Interests, according to Lasswell and Kaplan (1950, 17), are "a pattern of demand and its supporting expectations." *Common interests* are those that are widely shared within a community and demanded on behalf of the whole community. Safe drinking water, for instance, is a demand made by nearly all communities and supported by their expectation that they are entitled to a safe, healthy environment.

At a larger scale, all of humanity shares an interest in the sustainability of the environment and the future of the human enterprise. Human rights—freedom, self-determination, and dignity—are impossible without a secure natural resource base on which to build and maintain them (Sachs 1995). Modern science, including psychology, sociology, economics, political science, and policy sciences, supports the close interdependence between a rich natural environment and respect for human dignity. Policy-oriented professionals should thus be particularly interested in helping to clarify and secure communities' common interests, local or global. In fact, it is the work of professionals to solve policy problems in the common interest.

We all recognize, too, that there are special interests. *Special interests* benefit only part of a community at the expense of the rest of the community. There are many kinds of special interests, most of which mask their claims in the symbols and language of the common interest. For example, outfitters who guide hunters in Forest Service wilderness areas have a special interest when they demand a guaranteed number of state hunting permits, exclusive use of part of the national forest, or minimal interference in their hunting practices (for example, placing salt blocks just outside national park boundaries to bait game out of the park and into clients' gun sights). They may nevertheless justify their demands in terms of the common interest by identifying themselves as small businessmen victimized by state and federal bureaucracy, just trying to make a living for their families. They may feel hampered in their pursuit of their own versions of the American dream. But their special interest is easy to see through, regardless of how they try to justify their actions to the rest of the community. The demand for a monopoly economic market and exclusive use of public lands, while using questionable ethical practices (for example, salting), is clearly a special interest that benefits no one else.

Determining if a policy process seeks to clarify and secure common interests is not always easy, but it is possible to find common interests among conflicting special interests (figure 1.2). Doing so requires seeing what happened in implementation and how the overall process was carried out. Cromley (2001) lists three criteria for determining if a policy process serves common interests. First, is it inclusive and open to broad participation? Second, does it meet the valid expectations of participants? Third, as the policy is implemented or practically tested,

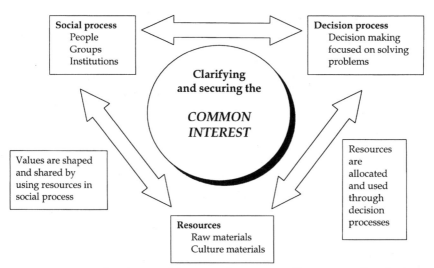

Fig. 1.2.   A generalized view of the natural resource policy process. People carry out decision processes in order to allocate and use resources, which affects how values are shaped and shared in society. The policy process is the means by which people clarify and secure their common interests.

is it responsive and adaptable in achieving the goals as the context changes?

Conclusions

Natural resource professionals—in forestry, range management, wild-life management, recreation, landscape architecture, fisheries, hydrology, and numerous other areas—are often more oriented toward "the resource" than toward the public policy aspects of their jobs. Yet practitioners in all these fields are faced with complex problems in dynamic settings, where the issues are partly scientific and partly not. Such problems, which demand responses but are not amenable to conventional scientific approaches, are often understandable and tractable using the policy sciences, a branch of knowledge devised to deal with just these kinds of situations. The very term *policy sciences* emphasizes the need to join our biggest and most important decisions—policy—to systematic, empirical inquiry—science, in the broadest sense—thus producing insight and improved judgment both for human freedom and for the sustainability of the natural environment. This book uses the theory and concepts of the policy sciences to explain the policy process and to guide professionals in dealing more effectively and more responsibly with natural resource problems.

SUGGESTIONS FOR FURTHER READING

Ascher, W. 1986. The evolution of the policy sciences: Understanding the rise and avoiding the fall. *Journal of Policy Analysis and Management* 5:367–73.

Brewer, G. D. 1981. Where the twain meet: Reconciling science and politics in analysis. *Policy Sciences* 13:269–79.

Brunner, R. D. 1996. Policy sciences. In *Social Science Encyclopedia.* 2d ed. London: Routledge

Brunner, R. D., and W. Ascher. 1992. Science and social responsibility. *Policy Sciences* 25:295–331.

Clark, T. W. 1997. Conservation biologists: Learning to be practical and effective. Pp. 575–98 in *Principles of conservation biology,* edited by G. K. Meffe and C. R. Carrol. Sunderland, Mass.: Sinauer Associates.

Lasswell, H. D. 1970. The emerging conception of the policy sciences. *Policy Sciences* 1:3–14.

———. 1971a. *A pre-view of policy sciences.* New York: American Elsevier.

2 Fundamentals

A View of Individuals and Society

It is the job of policy-oriented professionals to help people clarify and secure their common interests. As we discussed, policy is a process of problem solving among people with varying perspectives and interests, as true for natural resources as for any other issue. So it is with people that we begin our inquiry.

Complex policy issues bring to light the many demands that different groups of people make, their aspirations and hopes as well as their fears, their expectations, their personalities and identities, their needs, their strategies. For the professionals involved, however, basic concepts of what motivates individual and collective behavior are often assumed, the language to describe these things is lacking, and the possibility that people's interactions can be empirically studied is overlooked.

This chapter draws on a large body of research on human behavior to provide a brief, policy-relevant introduction to the personalities and identities of individuals, their affiliations in groups and societies of all kinds, the importance of myth in giving meaning to our world and in helping professionals understand the policy process, the maximization postulate and the concept of bounded rationality, values as the basic medium of exchange in all human interactions, and the principle of contextuality. These form the foundations for the rest of the book.

Individuals

We begin our study with people because policy is really about human behavior—individual and collective—whether we are interested in natural resource management or some other field. Policy processes involve the operations people carry out as well as patterns of attention, sentiment, interest, loyalty, and faith among individuals and groups.

All societies and policy processes are composed of people interacting. Understanding human behavior begins with the person, personality, and identity. Individuals are born with certain apparent predispositions, and their personalities and identities are organized around basic motives and major beliefs. Besides being biological entities, people are made up of a *subjectivity* that makes each a unique personality. We can observe directly physical activities such as people's writings, but their subjectivities can only be inferred from their observable behavior. People may observe their own subjectivities, however, and many people are self-reflective about what they think, feel, and do. People act in response to their subjectivities (also known as human nature or basic predispositions) as well as to environmental stimuli. This is significant for policy processes because people understand and interpret events, interactions with others, in fact, everything in the world around them based on their individual subjectivities. It is important for professionals to acknowledge that, for this reason, facts mean different things to different people (Berger and Luckman 1987).

One thing that distinguishes individuals is how they respond to new situations. When considering new alternatives or different perspectives, some people are open and flexible, whereas others are rigid and predisposed to maintain the status quo. This, too, is partly a function of personality. Failures or problems may heighten a person's receptivity toward new perspectives, operations, and practices. New alternatives may also arise from learning, that is, from new insights into self or environment, or from other people. The policy process is the vehicle by which societies adjust and adapt to new realities, and if people's perspectives are highly rigid, they will be unable to change to meet the demands of new situations.

A person is an individual with an ego and a social self. An *ego* is an person using symbols, that is, things that have meaning or significance in any sense. The most important symbols are linguistic, but there are other kinds as well, including visual ones that play a considerable role in politics, such as flags. Both subjective and symbolic factors figure in all political processes in fundamental ways of which many people are not aware. For example, Steven Primm (2000, 6) captures the power of symbol manipulation in an article about grizzly bears in the American West:

Grandstanding politicians use grizzlies as a potent symbol to rally conservative voters: they equate grizzlies with loss of resource-extraction jobs, reduced recreational access, and interference from "outside interests." There is a subtext here worth noting: for some, grizzlies symbolize a loss of power, a loss of opportunities to practice skills (e.g., logging), and a loss of esteemed positions (e.g., bread-winners, providers) in our culture. These losses exist independently of grizzly bears, but it is bears that provide a focal point for their discontent.

Identification is the process by which individuals come to see themselves as members of some aggregate or group. A uniformed biologist, for instance, identifies with an agency and with a profession. Identification is the chief mechanism for the creation of the political "we" that lies at the heart of all political processes as groups and societies interact, and it is in terms of the "we" that demands are made and justified. People make demands based on their values, and we can determine what people value by observing the nature of the acts they carry out; that is, values are expressed in concrete situations. People also have sentiments (feelings and emotions) and expectations that are often expressed in symbolic forms (see Goleman 1995). Identifications, demands, and expectations make up people's perspectives—a key variable in the policy process.

Groups and Society

In part because of shared identities, expectations, and demands, people often form groups and societies, which are two important units in the policy process. *Groups* are made up of individuals with a shared sense of identity, and people usually identify with many kinds of groups. Groups integrate diverse perspectives and operations, and they have cultures, or mores. They affect the values of their members and may also affect values within the larger society, depending on the group's mores as well as its permeability and circulation. They tend to evolve procedural norms or rules for behavior that members seek, live by, and come to expect (although some individuals may be poorly adapted to a group's norms). Every person and group, no matter how powerful or individualistic, has a tendency to conform to relevant group norms. Norms are contingent on the context; those concerning killing are different in domestic society and on the battlefield, for instance. Groups of various sizes and configurations interact, cooperate, compete, accommodate, or oppose one another. Every group has interests, some special, some general, organized for the satisfaction of group values. For example, professional societies are often assumed to have a general interest, whereas business lobbies have a special interest. As might be expected, groups, interests, and cultures tend to favor certain personality types.

Individuals are almost always part of a *society,* which is made up of groups and their cultures. Some sociologists think of society as "a large complex of human relationships, or to put it in more technical language, . . . a system of interaction" (Berger 1963, 26). People—by themselves, in groups, and in society—have interests (demand patterns with supporting expectations), faith (sentimentalized expectations), and loyalties (sentimentalized identifications or demands), all of which are involved in group and societal behavior. Although scholars differ on the nature of societies, their dynamics, and their relation to individuals, some suggest that society defines individuals but is in turn defined by them. Individuals thus collude and collaborate with society, and just as individuals are fragile biologically and psychologically, so too are societies fragile. They tend to remain static but are subject to change and even catastrophic collapse.

Individuals' collective acts support the edifice of society and in some instances change society for better or worse (Lasswell 1960a). Society has many rules of behavior, which in turn endlessly confirm society's structure and norms. It is possible, however, for people to withhold their confirmation of society, its structure, and norms, but usually at great personal cost. Because society is so hard to change, one way to try to do so is to organize into groups. Another way is to withdraw from society, turn inward, and withhold participation; detachment is a form of resistance. Sometimes countersocieties spring into being. In many societies worldwide, the dominant society contains many smaller countersocieties, some of which may seek to manipulate the policy process to their advantage. People are quite inventive in circumventing and subverting dominant societies, but the latter are very adept at maintaining themselves and the status quo, even in the face of huge pressures to change.

A society creates a system of civic and public order for itself (McDougal et al. 1981). *Public order* is the maintenance of order in society by government, that is, by allocating authority and control in stable patterns to restrict unauthorized coercion and violence. *Civic order* is the maintenance of order by civic bodies such as families or peer groups. Violations of public order are punished in major ways, whereas violations of civic order are punished in minor ways. Stable patterns of public and civic order derive from a society's institutions, which are, in fact, relatively stable patterns for the shaping and distribution of values. The family, the courts, and businesses are examples of institutions that stay more or less stable over time and determine how values are shaped and shared among members or relative to outsiders. Societies, institutions, and cultures are held together fundamentally because of sanctions—both positive and negative—which may be applied by individuals toward themselves, or, in many instances, by the "expectation of violence" that would result if civic or public order broke down.

Individuals, groups, and societies have different foci of attention. Two people may attach very different weights to certain factors in a problem. Humans react to only parts of their environment—the parts that impinge on their attention, whether consciously or not—usually on the basis of their predispositions (Smith and Berg 1987). Rational responses of any kind are impossible without a properly organized frame of attention. Education and political action help organize people's attention. In the arena of the courts, legislatures, and government agencies, practices are established to draw attention to certain rules and procedures. The rules of evidence in courts are designed to exclude some matters from the attention of the participants. Military intelligence branches are responsible for deciding what to bring to the attention of the leadership. The attention frame of a person or group leaves a heavy mark on perspectives (identity, expectations, and demands). Strongly biased persons often do not allow themselves to appreciate facts and symbols that are disturbing to their perspectives. We often say of such people that they simply cannot see the facts or that they refuse to see them, and they may appear irrational.

Myth

Myth, or shared belief, is the glue that holds groups and society together (Lippmann 1965; Patai 1972; May 1991; Slotkin 1992). Unlike the conventional definitions of *myth* as a legendary or traditional story or as a false or fictitious belief, the policy sciences use the anthropological or sociological concept of myth as a set of basic operating premises, belief systems, frames of reference, outlooks, worldviews, or paradigms. In functional terms, a myth is a "stable pattern of personal as well as group perspectives" (Lasswell and McDougal 1992, 353). The point is not that myths are true or false in empirical terms, but that myth is our way of understanding ourselves and the world (Bruner 1990; M. Lerner 1996). All matters of morality and justification are embedded in myth. Conflict in policy processes can often be attributed to participants drawing on different myths and justifying their special outlooks in terms of those myths (McIver 1947).

Myths are made up of three parts: doctrine, formula, and miranda. The *doctrine*, or basic premise, is often expressed in abstract or philosophical terms and functions to affirm the perspectives of the group. In the United States, for example, the Declaration of Independence, written in 1776, provides some of our national doctrine in the statement that "Governments [derive] their just powers from the consent of the governed." The *formula* consists of the prescriptive norms of conduct that must be followed, the "thou-shalt" and "thou-shalt-not" parts of the myth system. The United States Constitution and the Bill of Rights contain part of the main American political formula, and many laws have been passed to guide implementation of the doctrine. The *miranda* (or symbols) are the relatively popular manifestations or expressions of myth. Lore, stories, popular legends, poems, heroes, and other symbols are all miranda. In the environmental movement in the United States, stories about John Muir, Teddy Roosevelt, and Aldo Leopold are symbols. Many sub-myths exist within any one dominant myth. Science, for instance, is a sub-myth within Western society with its own distinct doctrine, formula, and miranda (see Brunner and Ascher 1992), as is capitalism or environmentalism (Lasswell and Kaplan 1950; Kuhn 1962, 1977).

Myth is a key concept in understanding and operating effectively in policy processes. According to Charles Taylor (1989), what locates people in their world, what makes their responses appropriate, and what makes something a fit object or value for them to seek is myth. It allows us to articulate meaningful expressions of self within a context by providing a "crucial set of qualitative distinctions" (19), which form the basis of our judgments about what is worth doing and why. These qualitative distinctions about life—and a definition of "a good life"—are possible only because of myth. For instance, billions of people have lived their lives believing that production and reproduction— in other words, work and family—are the main foci of the good life. Myths provide the background, backdrop, or frame of reference, whether explicit or implicit, for our rational, political, and moral judgments, intuitions, or reactions. Taylor indicates that operating within some myths demands that the nature of the myth itself be explicit, although this is not always the case. It is often up to historians, philosophers, anthropologists, or sociologists to attempt an explicit formulation of the goals, qualities, or ends by which people have lived and ought to live. Many people today in our own society are unconscious of their own myth and unable to articulate it.

Myths are created over time as people, groups, and societies give meaning to the experiences of their daily lives and their relations to other people and the natural world. In the modern world, besides elites, parents, and teachers, the mass media are particularly important in promoting and adjusting myths. Other factors are at play as well. For instance, in describing the significance of the frontier myth in American history, Richard Slotkin (1992, 8) points out that "the work of myth-making exists 'for the culture' that it serves, and we therefore speak of it as if it were somehow the property or production of the culture as a whole. But the actual work of making and transmitting myths is done by particular classes of persons; myth-making processes are therefore responsive to the politics of class difference."

When conditions and environmental relations change, the supporting myths also change. Although myths are generally maintained and updated adaptively, they sometimes become outdated and maladaptive so that mythic residues from past times may no longer explain and integrate people's current experiences meaningfully. In situations in which changing conditions require an adjustment in the prevailing myth, yet people who are well socialized to the myth resist change, problems may surface and tensions may rise. Many policy processes in natural resources today are about reconciling the beliefs of old and emerging myths. For example, in his book about the loss of biodiversity and environmental devastation in Australia, Tim Flannery (1994) concluded that the country's culture was ecologically maladaptive and posed two penetrating questions: "What laws and values should Australians encourage; and what artefacts and ceremonies should Australian society adopt in order to symbolise these goals?" (400). He captured in this single line the power, the meaning, and the need for myth—doctrine, formula, and miranda—and he understood that myth is both the cause of and an essential element in any solution to environmental problems.

Disenchantment in the modern world stems from the undermining of traditional myths (Lasch 1978, 1991; Berman 1981, 2000; Mander 1991). Old horizons and foundations are going or gone, and the world is now problematic. Old myths that seemed as unchangeable and as solid as the universe—in particular, those about how we use natural resources—are now disintegrating. A new situation exists in which the problem of meaning arises frequently in daily life. Because myths are changing, we are dealing with new interpretations and shifting perspectives that have huge consequences for the process of making meaning for ourselves about the world.

The Maximization Postulate and Bounded Rationality

The policy sciences view individual and collective behavior in terms of the *maximization postulate* (also called the optimization postulate), which says that "living forms are predisposed to complete acts in ways that are perceived to leave the actor better off than if he had completed them differently" (Lasswell 1971a, 16). This principle, which focuses on individuals' perception of acts open to them in given situations and of the advantages and disadvantages of each, says that people tend to behave in ways that they perceive will benefit them. The benefits may be psychological (that is, improving their mental well-being) and not evident to others. Human behavior is not predetermined, random, or necessarily rational but is very selective based on individual predispositions and environmental stimuli. It has a subjective component that may weigh benefits and costs of a particular act in rational or nonrational ways. It is a person's own subjective perceptions that matter.

Perception is a key element in the maximization postulate. According to Lasswell and Kaplan (1950, 70), "it is entirely possible for a person to be mistaken not only as to means for attaining his values, but even as to what his values are." The meanings and implications that people perceive or understand about themselves and their environments are largely subjective and may not correspond very closely to the external world. People tend to subordinate the external environment to their conscious or subconscious outlooks or predispositions and ignore or distort the external environment partially and selectively. In other words, they are *selectively attentive,* and this, unfortunately, leads to chronic misperception of events and processes. We all restrict or bound our attention to the external environment to some degree. There is simply too much happening to pay close and conscious attention to everything. No one is all-knowing, of course, and we all misperceive, although not to the same degree or in the same way. Each of us has a subjective map in our heads of our selves relative to the external environment. This map shapes how we understand other people and ourselves, how we act, and how we see the consequences of our behavior.

The concept of *bounded rationality,* as it was termed by Herbert Simon (1983, 1985), who won the Nobel Prize for his lifelong work on human reasoning, perception, and decision making, may help elucidate the maximization postulate. Simon explained that the environment in which we live is considered reducible into separate problems— many thousands of them. In order to make life practical, we simply ignore most environmental variables. If we did not, the world would become overwhelmingly detailed, complex, and confusing. Unfortunately, this leaves us open to ignoring key variables and misperceiving our environments and ourselves, with harmful consequences. Simon speculated on why evolution might, for a host of practical reasons, produce animals like us capable of bounded rationality (this property, to date at least, has shown selective advantage). He reviewed a great deal of psychological research that directly supports the idea of bounded rationality and the maximization postulate. An earlier formulation of the former is contained in Walter Lippmann's (1965) "The World Outside and the Pictures in Our Heads" (a chapter in his *Public Opinion*). In Lippmann's view, the world outside is so complex and we are so limited that "we have to reconstruct it on a simpler model before we can manage with it. To traverse the world men must have maps of the world. Their persistent difficulty is to secure maps" (11).

The maximization postulate is the most significant innovation in the contemporary study of social process (Lasswell and McDougal 1992). It is a practical tool for inquiry into any specific situation because it can be used to infer people's subjective map of self and situation based on direct, systematic, empirical observations of their choices and behavior (Brunner 1996a). Inferences about people's perspectives are necessary because the dynamics of their thinking are not directly accessible to outside observers (Lasswell 1960a), but in making inferences about people, we must describe their acts and the situation as comprehensively as possible within practical limits of time and resources.

Values

The basic medium of exchange in all human interactions is *values*, that is, the things and events in life that people desire, aim at, wish for, or demand (Lasswell 1971a). In all interactions, interpersonal or international, values of one sort or another are created, accumulated, given, received, taken away, lost, or in some other way transferred among people. This process of "shaping" and "sharing" values is fundamental in the policy sciences' functional understanding of social process. It allows participants—but, more particularly, analysts or observers of policy processes—to describe the myriad kinds, scales, units, and stuff of human interactions and transactions in comparable functional terms.

For instance, consider the case of a neighborhood affected by industrial pollution. It is easy in conventional terms to see that the large, powerful, wealthy, and well-organized industry gets its way while the individual residents who try to fight it by enlisting the help of grassroots groups and invoking antipollution regulations lose out. Looking at this case functionally, however, we can see that people and groups use the values they already have to acquire other values. The neighbors, for instance, draw on the affection, respect, and rectitude among themselves to band together, seek out the power, enlightenment (knowledge), and skill of organized groups, and invoke the power of the law to seek redress for the loss of well-being, power, respect, rectitude, and wealth that they suffer at the hands of the polluting industry. We can see also that policy processes have outcomes and effects on participants and society, such as the loss of trust in the government agency whose regulations are flouted or the discrediting of the grassroots group that fights the industry (losses of respect and power). These changes can be characterized in terms of which values, people, and institutions are left better off and which lose out. In other words, who gets what, when, and how.

The term *value* is used in the policy sciences without any metaphysical, ethical, or other connotation, so that this usage differs from its everyday meaning. A value is simply a desired object or situation, and the act of valuing something is valuation. In any society, institutions are specialized to particular values; institutions establish and maintain relatively stable patterns of value shaping and sharing. For instance, the family preserves patterns of affection, well-being, rectitude, respect, and so on, the institution of slavery perpetuates patterns of powerlessness, banks and financial organizations preserve certain patterns for the accumulation and distribution of wealth, and the health care system is specialized to the value of well-being. Because institutions significantly influence how values are shaped and shared in society, the term *value-institutions* is often used. Value-institutions serve to focus people's attention, organize their thinking about how the society works, and stabilize their expectations about how they will interact among themselves and with their environment. For example, we know how to operate within our own societies, but travel to another country usually forces us to spend considerable effort in figuring out how the value-institutions function so that we can go about meeting our basic daily needs and conducting our business. Having stabilized expectations about society and its value-institutions means that we all share socially valid, realistic expectations about the consequences of our behavior. It also brings some rationality to interpersonal and group interactions. Another way to state the social process framework, then, is to say that people pursue values through institutions, using resources that affect their environment.

The social process framework recognizes eight categories of values

(Chen 1989; Lasswell and McDougal 1992). There is no ranking of importance or preference among them.

Power refers to participation in decision making. People want to help make and influence the decisions that affect their own communities. Power is exercised in many forms, such as voting, decision making, organizing people into blocs, or deploying police or military forces for internal or external control. Elected and appointed officials in government are influential in community power relationships, as are many other kinds of professionals and leaders. Power institutions may include political parties, voting blocs, and elections.

Wealth refers to control of resources. It affects production, distribution, and consumption of goods and services in society. We all figure into the process of creating and distributing wealth, as do economists and business organizations, the active managers of wealth.

Enlightenment is the accumulation of knowledge (comprehension or understanding). Information (which consists of unorganized or unrelated facts) is gathered, processed, and disseminated in many ways. Scientists and educators in research organizations and universities specialize in enlightenment, as do the reporters and journalists in the mass media.

Skill refers to the acquisition and exercise of talents of all kinds—professional, vocational, or artistic. People may already possess skill or have latent talents that seek development. Organizations specializing in skill include labor unions, schools, and artistic cooperatives.

Well-being is safety, health, and comfort. Institutionally, medical professionals work in hospitals and related organizations to provide well-being, as do churches, social services, and other community help groups. Firefighters, police, and other community protection services, including the military, are also examples.

Affection is love, intimacy, friendship, loyalty, and positive sentiments among family, friends, and community. Organizations include family and friendship circles.

Respect is recognition, freedom of choice, and equality. Awards, honors, and everyday patterns of deference common in social interaction are practices associated with respect. Friends and colleagues and groups that present awards (such as the Nobel Prizes) are available to recognize outstanding people.

Rectitude is participation in forming and applying norms of responsible conduct. It is judgment in terms of responsible or ethical conduct. Rectitude is encouraged by religious or other ethical systems taught by churches, by families, or by others.

All the endless things people say they value can be functionally classified as one or more of these eight values. Based on decades' worth of its application in sociology and anthropology as well as in the policy sciences, we can depend on the exhaustiveness of this set of categories and on its utility for classifying the vast diversity of human values and for guiding our understanding of social process. But many people first react to the brevity of this list by demanding to know why it does not include freedom, justice, democracy, equality, and others of our most cherished ideals. Those in natural resource fields might ask about environmental protection or sustainability, for instance. In fact, they are all there. The call for freedom is simply a demand for self-determination—in other words, power and respect. Justice is an equal opportunity to pursue a good and right life—that is, to acquire adequate levels of all eight values. Sustainability is the maintenance of natural environments for well-being (that is, security and health) and—for those who believe that nature has rights of its own—rectitude.

The eight values are spelled out in the Universal Declaration of Human Rights and many other national and international declarations and constitutions. One example is the 1992 Declaration on Environment and Development produced by the United Nations Conference on Environment and Development held in Rio de Janeiro, Brazil. The declaration articulates a set of twenty-seven principles to guide decision makers and people everywhere to achieve worldwide sustainable development. It is a carefully expressed statement of world community intent. John Batt and David Short (1992–93) analyze the Rio Declaration to find out who gets (or does not get) what, from whom, how, and under what prevailing social conditions—in other words, how the eight values are to be shaped and shared in world social process. The very first principle declares that "human beings are at the center of concerns for sustainable development" and are entitled to good health and economic well-being. Principles 1, 3, 5, 6, 7, 8, 10, 13, 14, 15, 17, 18, 19, 23, 24, 25, and 26 all stress well-being, which is, in fact, the life-sustaining value that is at the heart of the document. Similarly, each of the eight values is addressed in one or more of the principles: Principle 3, for instance, speaks to affection, Principles 9 and 10 to enlightenment, and Principle 10 to respect; power, the ability to shape political outcomes, is covered by twenty of the twenty-seven principles. Batt and Short conclude by saying that the Rio Declaration "demonstrates a clear-cut preference in favor of human dignity, ecological maintenance, and an equitable worldwide distribution of the eight values" (292).

In most communities, whether global or local, the eight values are unevenly distributed among members (Lasswell 1951a). A group whose members have a similar share of one or more values is called a *class*, and usually three major classes are recognized along a continuum: elites possess a high proportion of values, mid-elites the next greatest amount, and rank-and-file the least. Individuals and groups may be members of the elite with respect to one value in one context yet members of the rank-and-file in terms of different values or the same values in different contexts. The concept of culture distinguishes communities from one another by their relatively unique patterns of community values and institutions. Different cultures seek different sets or mixes of values and use different institutional patterns to shape and share values. Within each culture there may be many subcultures, which, in policy processes, may represent special interests that conflict or cooperate in their pursuit of different values.

The Principle of Contextuality

Ethology, applied anthropology, and sociology all call for thorough contextual descriptions of behavior (for example, Alcock 1975). The policy sciences analytic framework and, in particular, social process mapping are helpful in this regard. The contexts that generate and condition all problems matter enormously in the public policy processes that must address those problems. The *principle of contextuality*—that all things are interconnected and that the meaning of anything depends on its context—applies here. Lewontin (1992) observed that it is not that the whole is more than the sum of its parts but that the properties of the parts cannot be understood except in the context of the whole (see also Dahlberg 1983). In Lasswell's view this is one of the most important lessons of modern historical and social scientific inquiry— that the significance of any detail depends on its connections or linkages with its context. And, as a result, to evaluate the role of any single institutional practice, for instance, requires an enormous effort in data gathering and theoretical analysis. The paramount importance of context demands that it be perceived by observers thoroughly and correctly. If the context is understood too narrowly, in a way that is too bounded, then significant details will be overlooked or misconstrued.

Much research in recent decades has corroborated the principle of contextuality. One important example is the work of C. S. Holling (1995), an eminent scientist and leading contributor to theories of adaptive management and ecosystem management for sustainable development. Holling notes that many environmental problems seem to be growing and that science as practiced acontextually is not solving them. This is largely because there are not only conflicting voices but also conflicting modes of inquiry and criteria for establishing the credibility of a line of argument. He writes that this "seems to define the conditions of gridlock and irretrievable resource collapse" (8). Holling has come to understand the importance of context in ecological management. He labels this a "systems view" of problems (we might call it a "policy-oriented" view), and he feels that the solutions, or improvements, lie in a new kind of science that integrates all the parts and is more relevant to the needs of policy and management. He notes that the premise of science using a systems view is that "knowledge of the system we deal with is always incomplete. Surprise is inevitable. Not only is the science incomplete, but the system itself is a moving target, evolving because of the impact of management and the progressive expansion of the scale of human influences on the planet" (13).

Hence, ecological scientists need to be fully contextual and adaptive. This is the rationale for "adaptive management" as promoted by Holling and his colleagues. Brunner (1997c, 1438) notes that Holling's insight "represents another important and independent adaptation of science to the requirements of management and policy decision in practice. Regardless of their disciplinary origins, such adaptations tend to converge on an outlook that is *problem-oriented*, . . . *contextual*, . . . and *multi-method*" (emphasis in original). In other words, these disciplines (and specifically the work of Holling) are moving toward the policy sciences. There are many examples of this shift under way in natural resource policy and management. Among recent examples of partial reinvention or convergence of the policy sciences framework are Bruce Byers, *Understanding and Influencing Behaviors in Conservation and Natural Resources Management* (1996), Gary Machlis and his colleagues, *The Human Ecosystem as an Organizing Concept in Ecosystem Management* (1997), Anne Schneider and Helen Ingram, *Policy Design for Democracy* (1997), Hanna Cortner and Margaret Moote, *The Politics of Ecosystem Management* (1998), and Richard Margoluis and Nick Salafsky, *Measures of Success: Designing, Managing, and Monitoring Conservation and Development Projects* (1998).

The maximization postulate, values, and the principle of contextuality all function as heuristics for us to learn about the social process, "intended to serve the function of directing the search for significant data, not of predicting what the data will be found to disclose" (Lasswell and Kaplan 1950, xxiii). For the professional who studies people and policy processes, these principles have implications, in particular epistemological and cognitive implications for what can be known.

Conclusions

Natural resource professionals' work must rest on a sound understanding of how not only natural systems but also human systems function—what motivates people's behavior, what nature means to them, the different ways they value resources, and the multitude (and importance) of linkages with each other and our environment. We cannot hope to participate in effective or responsible ways, nor can we possibly know where and how to intervene to improve natural resource management, without a grasp of these basics. The concepts described in this chapter are used throughout the book as we develop a policy orientation to the management and conservation of natural resources.

SUGGESTIONS FOR FURTHER READING

Lasswell, H. D., and A. Kaplan. 1950. *Power and society: A framework for political inquiry.* New Haven: Yale University Press.
Simon, H. A. 1983. *Reason in human affairs.* Stanford: Stanford University Press.
Slotkin, R. 1992. Conclusion: The crises of public myth. Pp. 624–60 in *Gunfighter nation: The myth of the frontier in twentieth-century America.* New York: Harper-Perennial.

3 Social Process
Mapping the Context

We all understand, of course, that people are involved in natural resource problems. It is often extremely difficult to analyze and solve the
problems because of the complex, competing interests of many participants with diverse perspectives, large geographic areas and long time
spans, and other factors. However we relate to natural resources, use,
misuse, exploitation, management, and conservation are all decidedly
human activities that serve varied needs for varied groups of people.
The interaction of every individual and organized interest in society—
in other words, the social process—constitutes the context of every resource problem, and neither the problems nor the decision-making
processes necessary to solve them can be understood unless their context is known. In fact, no problem exists without a context.

There is a way, however, part of the policy sciences framework, to
map a problem's context in order to make sense of the myriad particulars and to facilitate the search for solutions. It is a good way to gain
pertinent information about any social process, whether it is a job-
related problem in which you are professionally involved or whether
you are merely an observer of some issue of social import as reported
in the newspaper. It is simple yet sophisticated enough to be appropriate for both small and large situations. The policy sciences ask us to
see social process in functional terms: that all participants are seeking
values that they perceive will leave them better off, that they do this
through society's institutions, and that this process has identifiable outcomes and long-term effects on other people and on the environment.
This view allows us to sort out the complexity, make it manageable

and useful for decision making, and identify recurring patterns in the flow of events.

This chapter outlines a procedure for studying social process from the standpoint of policy-oriented inquiry. It clarifies what to look for, what to consider, and how to proceed in building your own contextual map, regardless of the issue.

Elements of Social Process

Regardless of the particular natural resource management issue you are considering, the practical challenge is to map social process accurately using the seven elements described below and summarized in table 3.1. Remember that this framework of categories is merely a guide to your inquiry (see fig. 1.1). It will help ensure that you have examined the context fully and have not missed or misinterpreted anything in your analysis (or someone else's analysis). It will also help you look at the interactions of people and institutions in meaningful, functional terms. When ranchers in the United States cry that raising grazing fees on public lands will put them out of business, for instance, you will be able to understand their demand in relation to their perspectives, their situations, their bases of power, and other factors. Fun-

Table 3.1. An Overview of the Social Process Categories and Some Questions to Ask in Mapping Them

Category	Questions to ask
Participants	Who is participating? Identify individuals, groups, and institutions. Who would you like to see participate? Who is demanding to participate? *Individuals* *Groups* *Organizations*
Perspectives	What are the perspectives of those who are participating? Of those you would like to see participate? Of those making demands to participate? What would you like their perspectives to be? Perspectives include: *Demands*, or what participants or potential participants want in terms of values or organization. *Expectations*, or the matter-of-fact assumptions of participants about past and future. *Identifications* (including myths—doctrine, formula, miranda or symbols), or on whose behalf are demands made?

(continued)

Table 3.1. continued

Category	Questions to ask
Situations	In what situations do participants interact? In what situations would you like to see them participate? *Ecological or geographic information* *Temporal dimension* *Institutionalization* *Crises or intercrises*
Base values	What assets or resources do participants use in their efforts to achieve their goals? All values, including authority, can be used as bases of power. What assets or resources would you like to see participants use to achieve their goals? *Power* is to make and carry out decisions. *Enlightenment* is to have knowledge. *Wealth* is to have money or its equivalent. *Well-being* is to have health, physical and psychic. *Skill* is to have special abilities. *Affection* is to have family, friends, and warm community relationships. *Respect* is to show and receive deference. *Rectitude* is to have ethical standards.
Strategies	What strategies do participants employ in their efforts to achieve their goals? What strategies would you like to see used by participants in pursuit of their goals? *Diplomatic* (negotiation) *Ideological* (ideas) *Economic* (goods) *Military* (arms)
Outcomes	What outcomes are achieved in the continuous flow of interactions among participants? Outcomes can be considered in terms of changes in the distribution of values. Who is indulged in terms of which values? Who is deprived in terms of which values? *Outcomes* also refers to the societal practices or institutions by which values are shaped and shared. How are practices changing? How would you like to see practices change? What is your preferred distribution of values? *Values accumulated or lost* *Decision choices*
Effects	What are the new value-institutions, if any? Are new practices put into place? Are old practices maintained? What forces promote new practices? What forces restrict new practices? *Values accumulated or lost* *Institutional practices* *Diffusion or restriction of innovations*

Source: Lasswell 1971a; Willard and Norchi 1993.

damentally, this framework will help you see that the social process engaged in any natural resource problem is not just a truckload of indecipherable and irrelevant details but that there are manageable factors at play that operate according to understandable forces. Moreover, you can collect reliable data about these variables to help communities and decision makers decide important issues.

Participants

Natural resource policy is both a process and a product of people interacting. As Lasswell and McDougal (1992, 591) warn, "Never . . . lose sight of the fact that human beings are involved." The first task in building a map of the social process in your particular case is to identify its participants, both individuals and groups. Groups may be either governmental (local, regional, national, or transnational) or nongovernmental (private associations, interest or pressure groups, or political parties). Participants may include those who are affected by the problem or situation at hand, those who are not currently part of the process but should be because of special contributions they could make, or those who are demanding to participate. All of these parties should be explicitly and systematically identified and described. Lasswell and McDougal (1992) recognize several kinds of participants: individuals, nations (territorial groups), intergovernmental and transnational groups, political parties and orders, pressure groups and gangs, and private associations. Another classification system identifies participants as experts, authorities, special interests, and the "unknowledgeable" (Thompson 1966).

Recognition of participants is a key early step in any policy process. Too often one group will try to control the process by controlling the selection of other participants. Some state game and fish departments in the American West, for example, have been known to claim that conservation groups do not have a legitimate right to participate in decision making concerning wildlife issues because they do not have direct or economic interests at stake or because they oppose hunting.

Perspectives

Each participant will likely see the policy problem at hand in a different way, so it is important to map participants' perspectives accurately in order to understand differences and similarities and find common interests. People's behavior and interpersonal interactions in social processes directly reflect their perspectives. Getting data on perspectives is challenging, but information can be obtained by observation, in interviews, and from the written record. Perspectives are made up of identity, expectations, and demands.

Identity lies at the heart of a participant's perspective. "We professionals" and "we conservationists" are expressions of identity. Even though individuals modify one another's behavior endlessly, each person shows a relatively stable pattern of behavior, which can be identified and studied to reveal his or her identity. Accordingly, people's personalities can be observed and described, and inferences can be made about them.

Identity is partly expressed in the symbols a person uses. People use words and other symbols to draw boundaries and show their relation or affiliation to other individuals or groups, that is, how they identify with society as a whole or with various institutions or groups. Every individual uses symbols of thought, expression, speech, dress, or action that refer to self and others, and this symbol usage can be recorded, described, and mapped. For example, men who want to be identified as westerners in the United States may display symbols of identity such as cowboy boots, large silver belt buckles, and wide-brimmed hats. Individuals, groups, and even nations all have multiple identities, and as a result, the identities in a social process are far more abundant than the actual number of actors or participants involved. Foresters may see themselves as foresters, professionals, employees, spouses, parents, community members, moral beings, public servants, and more. The symbolic identity divisions are almost infinite.

In addition to symbol usage, characteristic and stable patterns of practices demonstrate the identity of individuals and groups and reflect their values. For example, people oriented strongly toward wealth may spend money conspicuously and flaunt their possessions. Or academic societies may carry out educational programs to promote enlightenment. Entire societies may also exhibit particular personalities or identities. One may be aggressive and warlike, whereas another concentrates on knowledge and service.

Individuals and groups can be characterized as more or less parochial or universal in their identifications. Parochialism is a very narrow, close-to-home outlook on the world, whereas universalism is a broader, more encompassing view that takes into account the experiences of all humanity. Parochial people tend to identify themselves as living the good, moral life; they may take it for granted that their way of life is superior to all others. Their expectation regarding the intentions and capabilities of strangers is that they are probably malevolent and strong. This leads to the "us versus them" mentality so common in the world. In many conflicts over natural resource management, some participants hold clearly parochial views, whereas others seem to take more inclusive outlooks.

Parochialism has diminished somewhat over time, according to McDougal et al. (1988). Today, "the 'syndrome of parochialism' is seen for what [it] is—a barrier to the development of cooperative solutions to problems" (Batt and Short 1992–93, 250). Historically, the onset of urban culture millennia ago began a slow substitution of more universal outlooks for parochial ones. Universalizing religious doctrines were diffused widely over the course of centuries. The American and French Revolutions, and others that followed, dispersed doctrines that were legalistic and philosophic, also enabling more common, universal outlooks. These secular doctrines are behind the Western democracies and science of today (White 1967). Currently, globalization of the world economy functions to diffuse an even more universal identity to citizens of the world. As yet, however, there is no firm global system of public order because diverse parochial identities all around the world have prevented its development.

The maximization postulate and the principle of contextuality largely explain parochialism, which is a fundamental, universal human tendency. People operating with self-interest, bounded rationality, and selective inattention in local contexts dominated by subcultures and local myths will always manifest strongly parochial perspectives. Even with increased knowledge about global environmental destruction and the necessity of maintaining and restoring a clean environment for all life to enjoy, "the syndrome of parochialism has been continued into a world environment in which it has become species-destroying, not species-protecting" (Lasswell 1994, lxxxv). The parochialism of people's perspectives is a chief culprit in environmental destruction. Paradoxically, it may also be a salvation if sustainable development in innumerable local contexts can be made workable (Paehlke 1995; Scott 1998; see Chapter 8).

Identity is shaped by myth. Answers to the question "who am I?" may focus not on a name or genealogy but on one's commitments (the most important being on whose behalf values are sought) and judgments about what is good, valuable, and right—all of which are supplied by the myths within which one lives (Taylor 1989). Myth is what gives meaning and orientation to life's experiences, and lack of a myth can lead to severe forms of disorientation, or "identity crises." To have no identity means to be lost, without a frame or basis for creating meaning, a painful and frightening condition. At the higher level of society, myth is the total body of perspectives in a culture or subculture. "All social relations—the whole texture and the very being of society—are myth-sustained, and . . . all changes of the social structure are mothered and nurtured by appropriate new myths. Myth is the all-pervading atmosphere of society, the air it breathes" (MacIver 1947, 39).

Identity is closely connected to the expectations and demands people have about the world. Like identity, expectations are features inside a person. They can be inferred, in part through the demands that people make on one another. For example, a person's demands may include a strong desire, based on rectitude, to protect endangered species. Such a person has expectations that species will continue to decline, but neither his identity nor his expectations become apparent until he acts by making demands on other people to stop logging or development to save species. Demands are observable, and we can use them to make inferences about a person or a group's identity and expectations.

Expectations, or what people regard as likely to happen in a social process, can be articulated in terms of the eight value categories. Positive, or optimistic, expectations are held by people who anticipate being "indulged" in a certain value (say, respect)—that is, that they will gain new respect or at least not lose it. Negative, or pessimistic, expectations are that they will be "deprived" of respect, or that they will lose it or be denied. People's *demands* are also about value indulgences or deprivations, but with the additional element of preferences about practices: people demand that practices change to indulge them in certain values. For example, if people demand knowledge, then the practices involved in education are sought; if power is the value demanded, then the practices involved in power sharing are sought. Demands are made almost endlessly throughout social process by individuals, associations, groups, parties, organizations, and nations. Demands are constantly being made on oneself as well as on others. Individuals impose demands in terms of success in life (enjoyment of all eight values), for instance; a nation demands of itself that all its citizens be protected.

Perspective, then, is a key feature of people interacting in social process. We cannot understand natural resource policy and management without a good understanding of people's identities, expectations, and demands.

Situation

We also need to know about the situations in which participants—given their differing perspectives—make value demands on one another and affect and use the environment. The situations in which people interact vary over time as people's values change; people's values determine the significance of situations. Situation refers to the "zones" in which social interactions take place. There are four dimensions that characterize situations. First, the ecological or geographic dimension accounts for the spatial scale and related features in the area of concern. This is, of course, one dimension that resource professionals already appreciate through their work on ecosystems, habitats, niches, and geographic information systems, for instance. In this context, we need to add an understanding of the spatial components of human activities. Second, the temporal dimension is the timing of events and processes in the situation at the focus of attention. Different aspects of situations operate at different rates. In natural systems, disturbance events vary over time, as do the rates of human social interactions. These must be described and understood to the extent possible.

Since, again, institutions are specialized to particular values, the third aspect, institutionalization, refers to the structure of how values are allocated in particular contexts. For example, power may be centralized or decentralized or concentrated or dispersed. Regimentation and bureaucratization may be increasing or decreasing. The management of national parks in the United States, for example, is centralized, bureaucratic, and somewhat regimented, but management of ecosystems is often decentralized, informal, and individuated. The degree and type of institutionalization has significant consequences in policy processes.

The degree of organization of a situation may be especially important. Political arenas simply do not exist in some instances. For example, just a few decades ago there was no organized arena to conserve endangered species, but today, with a social process established and systematized to support the Endangered Species Act, in addition to many state and local laws and regulations, the arena addressing this issue is relatively well organized. An underorganized situation currently exists in the Greater Yellowstone region as the national parks and surrounding regions attempt to address the consequences of mushrooming tourism and development (Clark and Minta 1994; Clark 1999). Calls for greater ecosystem alliances and greater coordination are demands for more organization in this arena.

Institutionalization can also be considered in terms of its plurality or singularity. If a forest plan review is truly open to input by people with diverse values, then the situation is plural. If a grazing lease is to be let on public lands and only the leasing agency's district manager and the rancher are effectively involved in deliberations, then the situation is singular. Keep in mind that in all situations, all values are always at stake although some may be much more prominent than others. A true democracy always attends to all eight values, to be shaped and shared widely by diverse people.

The fourth important dimension of a situation is whether it is in crisis. People behave quite differently in crises, whether they are ecological (Chernobyl or Bhopal), social (race riots), political (coup d'état), international (war), or otherwise. Both perspectives and practices may change during crises, and as a matter of course, people, institutions, groups, cultures, and societies tend to try to avoid crises. Good policy works to do just that, but sometimes crises present themselves and must be addressed.

Base Values

People not only seek the eight values as goals; they also have values at their disposal to use in social and policy processes. In any social process, individual and institutional participants have value assets and liabilities in terms of the goals or objectives they seek. The values already possessed by people or groups are called *base values,* which can be drawn on or employed in many ways as a means to get more values. When values are demanded or sought as ends or outcomes they are called *scope values* (scope, in this sense, meaning "aim" or "purpose"). For example, the National Oceanic and Atmospheric Administration uses its knowledge and skill to study, monitor, and forecast weather, and in turn, these base values are used to seek scope values such as more financial support for the agency, greater respect, more enlightenment, and power. At an individual level, Mahatma Gandhi used the rectitude in world opinion, the affection of India's people, and worldwide respect for his nonviolent methods to end Britain's colonial rule of India and gain power for the nation's people.

Power is an especially important base value that may be used for the acquisition of more power or as a means for acquiring each of the other seven values. Suppose the head of a state natural resource agency uses the high-level position to secure appointment to the planning committee of an international conservation organization, thus using power as a base for securing more power. Power may also serve as a base for enlightenment in that more information is likely to be available to those in high positions. Power also gives the individual opportunities to exercise rectitude to improve resource management and thereby gain respect as well. People in power also generally command higher salaries and other financial benefits, and it is easy to see that power and wealth also characteristically lead to improved well-being, both mental and physical. Gains in all these values may result in invitations to the individual to speak at public and private meetings, which garner even greater skill, respect, and affection. Wealth is another base value that is exceptional in its ability to "buy" other values. Power and wealth are widely (and often cynically) recognized for their predominant influence in our society. Nonetheless, all values are used as bases to achieve goals.

The eight classes of values affect one another in endless and complex ways as people use base values to achieve scope values. Although we have used largely individual examples, these concepts apply to collective action as well, including activities by families, neighbors, groups of all types, organizations, communities, and nations. It should be clear that the value categories and the concept of "value-institution transactions" is a particularly powerful and functional way to describe all human activities in social process.

Strategies

People employ various strategies to pursue their scope values. Lasswell (1971a, 26) defined strategies as "the management of base values to affect value outcomes." There are four basic strategies for the manipulation of base values. *Diplomatic strategies* are practices that use communication among leaders or elites of any group, government or nongovernment. For example, negotiations between lawyers representing different clients in a lawsuit represent diplomacy. Discussions between leaders of business and conservation groups in an environmental dispute also constitute a diplomatic approach. *Ideological strategies* are practices that involve communications to a wider public. Public talks, newspapers, and white papers for broad dissemination are examples. Propaganda is an ideological strategy. A lawyer arguing a case before a judge and jury uses an ideological strategy, as does an environmentalist issuing a press release or speaking at a public hearing. These two strategies—diplomatic and ideological—rely heavily on rectitude, skill, respect, enlightenment, and affection as base values. Both are communicative in that they specialize in the use of symbols.

On the other hand are two different strategies that rely more heavily on the collaborative use of resources. *Economic strategies* are practices that rely on the production and distribution of goods and services. Boycotts are a classic example: an environmental group may encourage consumers to stop buying the products of a firm that carries out harmful practices, with the expectation that the loss of sales will hurt the company enough to force it to change its operations. *Military strategies* are practices that use resources as weapons. Attacks on other groups or countries are designed to deny them their livelihood, which is really depriving them of power, well-being, wealth, and so on. Specialists include armies and police forces.

In the natural resource arena, a conservation organization uses an ideological strategy when it writes editorials opposing a housing development that threatens to reduce critical wildlife habitat. It uses a diplomatic strategy when its leaders meet with the developers or the town leaders. It uses an economic strategy when it boycotts products or services of the developers. Rarely in this country do conservation groups resort to military strategies, although people who spike standing trees that are soon to be cut and milled (or "militias" who blow up federal offices or property) are employing a military strategy. These four strategies can be used collaboratively and persuasively or coercively.

In your analytical capacity as an observer or participant in policy processes, you should look at which strategies are used by which participants toward what ends and assess whether they are appropriate and effective. You may be able to identify potential strategies that have been overlooked or, within each strategy, practices that could be added or dropped to make each participant more effective.

Outcomes

Social processes have outcomes, results, or consequences that are measured in terms of values. Outcomes are the short-term, culminating events that indulge or deprive participants in a given situation. Outcomes are either desirable or undesirable, depending on the perspective of the participant. Winning an award is an outcome in terms of the value of respect; receiving a paycheck is an outcome in terms of wealth. Even though a single value may dominate, all values are nearly always at play at some level in social process.

As social interactions take place, decisions are made, practices are carried out, or institutions are changed, some participants are necessarily indulged in some values and some are deprived of some values in the outcomes. This is what is meant by the shaping and sharing of values through social process. In resource management, a plan is adopted, management actions are implemented, or an evaluation is made—all are outcomes that favor some people or groups over others. Some people and groups end up with more power, enlightenment, wealth, or so on than when they started, whereas some others end up with less. Outcomes also influence future social process. As you examine a particular situation, look at the outcomes value by value and ask how each value changed for each participant in the interactions that have taken place. As a professional with the responsibility to manage resources well, you should continually ask (in advance of any decision, during the decision-making process, and afterwards) what value outcomes each participant seeks and which ones each ends up with. Ask how each value might change for each participant if different sets of practices, actions, decisions, or scenarios were to take place. Ask what institutions and what practices are being advanced and which are being set back.

Effects

Effects are long-term outcomes in terms of values, institutions, and society. Effects are the net growth or decline of the values involved as a result of all the value outcomes. Effects are largely evidenced in institutional terms, and analysis of the kinds and abundance of organizations dedicated to the cultivation of particular values permits us to determine value outcomes in society (Lasswell and Fox 1979). We all recognize that over time changes in institutions and their practices occur as a result of social process and that these represent changes in the values that society holds dear. For instance, since the founding of the United States, there has been an increase in the number of schools and number of students in attendance. This has had the effect of increasing enlightenment. It has also had the effect, as supporters throughout history intended, of increasing both power through democratic participation and loyalty to the national government. The quality and quantity of government and other practices and institutions changed as an effect of social process. The policy sciences can help track change and monitor its impact on the achievement of goals.

As organizations appear, disappear, grow, or decline, practices in society change, sometimes drastically. In recent decades, for example, many nongovernmental organizations have sprung up, such as the Rocky Mountain Elk Foundation, whose practices of buying habitat, organizing members, and other activities have changed the face of wildlife conservation. Many federal agencies that deal with natural resources, especially the Forest Service, have undergone dramatic change. *Innovations*—new practices and value-institutions—may come about, and the processes by which they are diffused or restricted are important indicators of social process. Innovative change is a major source of cultural evolution. Some innovations are widely diffused through society and the world. Communism and its practices, for instance, saw a broad diffusion and then a rapid collapse and restriction in the twentieth century. Ecosystem management appeared in the last two decades of the twentieth century; usually framed as reform, it is an innovation in land management and policy (Grumbine 1990, 1994; Gunderson et al. 1995; for the Greater Yellowstone ecosystem, see Clark and Minta 1994). The new practices that it spawned have been partially incorporated and partially rejected—two methods by which diffusion and restriction of innovations take place. A history or trend map of resource policy in a country or region would likely show a convoluted sequence of innovation, diffusion, and restriction of varying institutional permanence and significance.

To determine effects, look at the degree of change in social process and, at the societal level, how institutions change over time. Fundamental changes may occur in the shaping and sharing of value-institutions. If a society is flexible, then change is likely to be gradual, but if it is rigid, change may come dramatically. It is desirable that change take place at a comfortable pace because rapid or revolutionary change may destroy people, values, and organizations.

Mapping Social Process

The professional's job is to use observation and other research to get empirical data on each element for particular situations.

A Method for Mapping Social Process

You can gather data on social process—participants and their perspectives, situations, base values, strategies, and outcomes and effects—from reports, interviews, publications, newspapers, libraries, conversations, questionnaires, letters, public and private meetings, and your own observations and research. Table 3.1 provides a series of questions for you to ask about basic components of any social process. Answer these as fully as you can given the purposes of your analysis.

Mapping social process thoroughly may seem like a straightforward task at first, but it requires work, rigor, and insight. To begin, it may be difficult to determine who the participants are in a given case. Clearly some people and organizations are directly involved, but others are increasingly peripheral. Where are lines to be drawn about who is in and who is out? The categories or kinds of participants listed previously and in table 3.1 can guide your initial mapping exercise. Think comprehensively at first about participants, and as you proceed in your analysis and learn more, think more selectively.

The term *stakeholder* is commonly used today as a label for people involved in a given case. This term is usually used in a narrow sense to include people who represent others (for example, representatives of business) and people who have direct (often economic) interests in a particular outcome. This definition excludes many kinds of participants who have other interests (for example, well-being, enlightenment, skill, or rectitude) as well as more general interests in the outcomes. Always be on the lookout for people and organizations that might be left out of efforts to address problems. If a policy process does not include and consider the demands of the right set of participants, ultimately, it is not likely to solve the problems fully and may create additional ones. Be as inclusive as possible, and beware of those who would narrowly limit participation as a way of furthering their own interests or increasing their own power.

Mapping people's perspectives is also challenging, but it is a key task that will give you broad insights into people's actions in the unfolding of events. Determining the demands of a person or group is relatively easy. It is usually a matter of public record what people say they want and how they act to get it. Compare what they say with what they do, and think carefully about how to account for any discrepancies. Mapping people's identities and expectations is more difficult. You can only make inferences about these from their observed behavior.

Delimiting the situation (or arena) of social interaction in terms of its geographic, ecological, temporal, institutional, and crisis dimensions requires both social and natural science research. The situational boundaries of a social process may not be easy to describe. One situation may grade into others, or it may change over the lifetime of the social process, becoming larger or smaller, more or less organized, and developing crises from time to time. Professionals trained in biophysical sciences may tend to focus on the geographic or ecological aspects of the situation and neglect other aspects. Again, beware of those who would define the situation for the advantage of special interests.

You can find out about the base values of participants simply by asking them directly, by observing their actions and choices, and by asking other people about the participants' behavior (Lasswell 1971a). People often have values, such as well-being, affection, and respect, that they fail to appreciate and put to good use in policy processes. Of all the value categories, respect is perhaps the least appreciated for its overall importance in social process. As you map base values and how they are used, remember that people rarely think and talk in terms of the eight categories we have defined. As a result, they will not be able to tell you directly about their value holdings and how they use them in social process. You must listen and observe carefully to assess what values are at stake and how they are being shaped and shared.

The strategies people employ, often a combination of the four, are usually clear, but they may change over time as one strategy is shown to be less effective than expected. Your job is to determine how people mobilize their base values using the strategies. You may be able to judge the effectiveness of different strategies if your research is well organized. Often in natural resource cases, participants use an ideological strategy, believing that if they educated everyone else to their point of view, problems would be solved. Not only does this approach avoid dealing with the real and legitimate differences among people, it does not address many practical and moral concerns, such as who is going to educate whom, about what, and how.

The outcomes and effects of social process, described in terms of values and value-institutions, can be determined while the process is underway as well as afterwards. Ask yourself: Who has ended up with more power? Less power? More respect? Less respect? Whose sense of rectitude has been reinforced? Whose has been undermined? Do these outcomes serve the common interest of the community? Do they help further widespread human dignity? What new institutional practices have been put in place? Have new programs been started or old ones shut down? With what effects in terms of how values are shaped and shared?

Knowing about the social process categories lets you quickly focus your research and decide what information is known or readily available and what must be searched out. Your study's boundaries are not pre-established but will emerge in the course of your investigation. Your approach should generally follow a case study format, which can best capture the multiple realities at play in complex human interactions (Yin 1989). Proceed in a grounded, emergent way (that is, through induction), rather than with a preset method or explanatory theory. Although many common social science methods may be unknown to professionals trained in the natural sciences, as the importance of social, economic, historical, psychological, or organizational factors becomes clearer, you must draw on modern social science research with appropriate standards and reliable methods. Any social or biophysical method, for that matter, can be used to locate and map social process data as the context demands (for example, Janesick 1994; Marius 1995). You might choose to develop your own social science knowledge and skills, or you might hire or consult with a social scientist to help you. Some organizations may benefit from adding a social scientist to their staffs.

The social sciences use powerful, empirically grounded research methods, ranging from positivistic approaches like those in the biophysical sciences to descriptive approaches similar to some naturalistic methods used in ecology and natural history studies. Descriptive, qualitative methods "investigate human behavior in its natural and unique contexts and settings by avoiding the artificial constraints of control and manipulation" (Isaac and Michael 1995, 218). This approach examines human behavior in real situations, relies on observational techniques, adapts itself to multiple circumstances, and recognizes both intuitive and explicit knowledge (Scott 1998). Because this kind of research studies human perception and multiple realities, often for applied purposes, it is little concerned with creating a final, unified system of knowledge or a grand theory.

In addition to using descriptive and experimental methods to determine causes and effects where appropriate, the social sciences make inferences about human behavior based on observation. Making inferences from data is an important function of research; its aim is coherence. Most people assume an ability to make correct inferences, and in our daily lives we make many inferences by recollecting experiences and using them to interpret present situations (Marius 1995). Without inference, we would have to reinvent life anew each day. Social scientists, as well as biological scientists, infer answers to some scientific questions. In doing so we strive to make sense of a behavior or situation and decide what it is and whether our interpretation is reliable. Researchers use inference to fill in gaps, round out, or complete a picture of a situation or event. Statistics can be a valuable quantitative method in this regard, but they require interpretation. By themselves, statistics tell us little, but what we infer from them can tell us a great deal. The use of social science methods in natural resource policy and management is increasing, but they have yet to be applied in ways that demonstrate their full potential. Chapter 7 describes some useful policy sciences methods.

Learning about social process and the methods required can enhance your personal understanding of policy processes and improve your skills in participating more effectively and more responsibly. You will, for example, be able to argue for the inclusion of some previously excluded group, speak up for the values of those who are disenfranchised, help balance the power holdings among participants, or see through the "common interest" appeals of special interest groups. You will be able to make projections about outcomes and long-term effects in terms of changes in institutions, practices, and values. Map the social process as fully as you can, keeping in mind the overriding professional goals of serving common interests, promoting human dignity, and improving the sustainability of the world's natural resources. Compare your map with others' to confirm the empirical basis for your findings. As the policy process proceeds and as you observe how decisions are made and come to understand how problems are defined and addressed, continue to update and refine your map of social process.

Two cases described below illustrate the social process categories and the analyst's job of mapping them empirically in order to understand the functional significance of social interactions. The first centers on a timber sale in the Medicine Bow National Forest. The second is a review of a massive effort undertaken by scholars at the Yale Law School to map the world community as a single planetary social process. It is an excellent example of the scope and utility of the policy sciences.

Logging in the American West

To give you some idea of how to put your social process mapping skills to use, I will describe the Tie Camp timber sale in the Sierra Madre Mountains of the Medicine Bow National Forest in Wyoming and Colorado in the late 1990s. I do not fully map this case but use it to illustrate how to develop a contextual understanding of a natural resource problem. Think of the interactions in the case in terms of a value-shaping and -sharing process that has outcomes and effects. Although newspaper articles do not offer the kind of analytic depth necessary to map social process accurately and completely, they can give you important information.

Perusal of several articles and letters in Casper and Laramie newspapers from 1997 to 1999 indicates that this proposed sale of 12.7 million board feet of timber stirred the public like none other in this forest's history. The proposed timber sale of 1,662 acres, straddling the state line, allowed 127 timber cuts, 80 of them clearcuts, in a roadless area bordering the Encampment River Wilderness Area. The forest had already been heavily clearcut. Twenty-five miles of new roads would be required at a public cost of $500,000.

We first need to identify the individual and organizational actors. The key participants in this case were foresters, conservationists, citizens, and the timber industry. The Medicine Bow National Forest was a major player, with the forest supervisor and certain district rangers playing particularly important roles. The Wyoming Game and Fish Department was concerned about effects of the timber sale on wildlife. Friends of the Bow/Biodiversity Associates, an environmental group, opposed the sale. The Western Forest Industry Council and the Intermountain Forest Industry Association spoke for the timber industry. Four mills in surrounding communities were served by timber from this forest. Not as well represented in the newspaper articles were the many people whose lives would be affected by the decision whether to cut the timber and how they might be affected. Our job as analysts is to identify all the people, groups, and organizations central and peripheral to the issue and to determine how the outcomes of this decision might deprive or indulge each with respect to certain values. It is especially important to try to identify "pockets of discontent" and people whose values might not otherwise be considered in the policy process. As part of its planning efforts, the Forest Service (or other organizations) could collect empirical data about these important variables just as it currently studies water yield or wood fiber production as related to various timber harvest methods.

Perspectives can be determined by what participants say and do. Words and actions suggest a pattern of demands being made by different participants and permit us to make inferences about their identities and expectations. The participants in this case were organized around two major perspectives. "Pro-timber" groups believed that the sale was essential for the forest's health and the local economy, while "pro-conservation" groups believed that the sale would overextend the forest ecologically. Nearly six hundred letters from the public were equally divided between these two positions. For instance, the Biodiversity Associates complained that they had been unfairly portrayed in the media as radicals, suggesting that they were feeling a loss of respect and affection. Yet, in the same letter, they accused timber of being a doomed industry concerned only with profits, showing the same disrespect they suffered themselves. The increasing intensity of demands from several sides suggests that the NEPA (National Environmental Policy Act) process in this case did not satisfy many people's demands for value indulgences.

As is usually the case, each of the primary participants billed itself as the true representative of the common interest. The conservation group argued that it had strong mainstream support in fighting for "the rights of the public to enjoy the natural heritage of land they themselves own." Industry contended that it was supplying society's demand for wood products in environmentally regulated ways—better from a global perspective than the devastation caused by logging in other countries. The Forest Service saw itself caught in the middle. The supervisor wrote repeatedly of seeking "more interaction, communication and understanding," of "a public collaborative process . . . to find 'common ground,'" of balance, of his agency's mandate to "ensure the best stewardship . . . while facilitating multiple uses." As analysts, we can collect additional data on these variables through empirical research, from which we can build a more functional understanding of the value-institutional significance of the actions and interactions of each group and of how each contributed to clarifying and securing common interests. As we are all well aware and as this case alone well illustrates, clarifying common interests is not an easy task for the participants or any would-be analyst.

We also need to know in realistic detail about the situation or zone of interaction in which this timber sale was being considered. Geographic, ecological, and temporal components of the situation provided much of the arena in which the policy process was focused. The situation clearly included concerns and impacts beyond the immediate forested areas for surrounding communities, the region, and the larger debate in the West over public lands. It included diverse time scales, from forest planning for periods of ten to fifteen years to recovery of logged areas measured in decades to potential long-term changes in value-institutional configurations. The newspaper articles relate that the arena was well organized, or socially structured; a number of clearly defined groups were participating in patterns of conflicting value demands that had been repeated in numerous previous cases challenging Forest Service timber sales under the NEPA process. It was also highly institutionalized; both the agency management of large areas of public land and the industrial use of forests were well-established patterns of practice. The rising intensity of debate about management of natural resources in the West also constituted part of the situation. Additional characterization of the situation will add to our developing picture of the social context surrounding the Tie Camp timber sale.

The base values brought to bear in the case are also illustrated in newspaper articles. I have already mentioned that some participants used rectitude—the conviction that they were doing the good and right thing—as well as the respect (and probably the affection) of at least some portion of the community to argue for their demands. Enlightenment (or knowledge, which, in this case, is largely biological science) was used by the Forest Service, conservationists, and the timber industry in different ways. Timber interests clearly had the wealth to establish industry associations for promotion, lobbying, and research purposes. The Forest Service also deployed skill in managing the forest as a base value. Of course, the federal agency also wields tremendous power in decisions such as this. These, of course, are only the most obvious uses of base values, but they reinforce the point that there is more to problem-solving than dueling scientists, wealthy industries, and powerful government agencies. As in all cases, all eight values— power, wealth, enlightenment, skill, respect, affection, well-being, and rectitude—were to some degree involved here. The analytic task is to find how they were manifest, how they were used, whom they served, how they interacted, who was left better off (and in what ways) and who was worse off, and how society's institutions and practices were changed.

The use of base values reminds us to characterize the strategies employed by the different participants. The letters to the editor, public meetings, mass appeals, dissemination of technical data, and economic and ethical arguments, for instance, were clearly ideological approaches. It is likely, too, that diplomatic efforts were also going on as leaders of various groups met to work out differences or exchange information. Economic strategies were in use as well.

The outcome, as yet undecided, will eventually be a decision by Medicine Bow National Forest. In value terms, it will indulge certain participants, reinforce their perspectives and value outlooks, and validate their strategies. Other participants will be deprived. Certainly, most of the participants have speculated on the consequences of choosing one alternative logging plan over another and what it might mean from their perspective.

It is too soon to determine the long-term effects of this social process on Forest Service practices. Medicine Bow National Forest may well change its practices as a result of this experience. Whether the forest innovates to improve social and decision processes or at least seeks to avoid or minimize problems for itself remains to be seen. If it does institute new practices, these may be used broadly or remain quite limited.

To understand the social process in this case fully would require additional research on most of the social process categories. Social science research can and should be conducted. Analysis of the data set will elucidate participants' perspectives, in particular, their patterns of demands. The decision-making process must respond to these demands. The framework can help you ferret out what information is missing from a policy process and what might serve as a debating point for finding common ground. When you understand social process in terms of perspectives, interests, and values as well as the other categories, you might then be in a position to help participants find an integrative solution, a "win-win" outcome in which value demands can be satisfied substantially for all or most of the participants.

World Social Process

Perhaps the most ambitious social process mapping effort ever made was carried out by Myres McDougal, Michael Reisman, and Andrew Willard (1988), who with their colleague Harold Lasswell mapped the entire world community as a single planetary social process. This effort, which took years to complete, covers all of human history, focusing on the past few hundred years. Their characterization of the world community is necessarily general but is supported by extensive detail in footnotes and citations, and it makes for fascinating reading.

According to McDougal and his colleagues, the world social process is composed of all the interrelated communities of the world. The term *community* refers to people interacting in patterns in which, while pursuing their own values, they must also take into account the values of others. This is what makes a community and leads people to engage in policy-making activities in order to clarify and secure whatever interests they may have in common. Once people realize that they are part of the same community, the chances that they will actively seek their common interests are likely to increase. This, in turn, leads to a system of public order, vests the community itself with the authority to make decisions, and gives it effective power.

The authors use the same framework of social process to describe the world community that was outlined in this chapter: people strive to maximize values using institutions affecting resources. They use the same set of elements—participants, perspectives, situations, base values, strategies, outcomes, and effects—and the same set of values—power, wealth, enlightenment, skill, well-being, affection, respect, and rectitude—that I have described. In particular, they look at how the outcomes of social process shape and share values and how people are indulged or deprived.

McDougal and his colleagues identify six kinds of participants in the world social process: individuals and five kinds of groups. The groups are territorial communities (nation states), intergovernmental and transnational organizations (for example, the United Nations), transnational political parties and orders (for example, communism), transnational pressure groups and gangs (for example, organized crime), and transnational private (or official) associations oriented toward values other than power (for example, fraternal organizations).

They describe trends in participation shown by each of these groups in the world social process, including, first, a demographic explosion since the fifteenth century that has increased the number of individual and group participants in social process, and, second, increasing internationalization.

Regarding perspectives, the authors note a major trend in today's world community—a growing similarity of perspective. As part of this trend, some people are developing an awareness of and inclusive identification with a single environmental system. Growing global interdependence means that the activities of any one subcommunity influence, and in turn are influenced by, the entire community. The overriding demand everywhere is for greater production and wider distribution of all eight values. The demand for human rights everywhere is just one example. Another trend is that the forceful diffusion of Western, science-based perspectives and operations has suppressed or replaced many perspectives worldwide, but in response some local communities are reasserting their historic identities, organizing their expectations accordingly, and demanding independence, self-expression, and control over decision outcomes.

Situations around the world are highly diverse and constantly changing, including the ecological dimensions that affect life itself and the continual rise of crises of all kinds.

An inventory of the base values (including resources) held by various participants is essential for understanding the ongoing "who gets what" of social process. Resources include aspects of the physical environment used according to their value potential; they change constantly depending on technological developments and other factors. Modern science and technology have permitted additions to the known stock of resources (in fact, some people believe that modern science ensures an unlimited resource base) but have also aided in their rapid depletion. Industrialized civilization is using the environment intensely and causing environmental deterioration or transformation of parts of the environment for nearly all human communities. Fortunately, many people now see a unitary global environment, and continuing discussion of sustainable development suggests that the world community is interested in changing the ways in which values are shaped and shared.

McDougal et al. indicate that the current world situation shows a complex and vigorous mix of negotiation, ideas, goods, and arms, which correspond, respectively, to the diplomatic, ideological, economic, and military strategies detailed in this chapter.

Outcomes of the world social process are constantly changing. In general, though, the past few hundred years appear to have produced, as a result of the universalizing forces of science and technology, a continuously increasing production and wider sharing of all eight values. Despite this, many people remain disadvantaged, especially in Africa, parts of Asia, and South America. Although it is difficult to measure outcomes in terms of the eight value categories, general characterizations based on a comprehensive and detailed review of world history indicate that all values are in flux.

The way in which future outcomes might unfold will be partly a continuation of long-term trends. The magnitude of value accumulation has increased over the past century, as has the trend toward wider sharing of values. Democratization in many communities is but one example. Again, this pattern has been shaped by the spread of science and technology. The authors expect this trend to continue for some time as innovation and diffusion accelerate, even in the face of restrictive forces such as fundamentalist religions. Today, there is clearly a growing and powerful demand for participation in the shaping and sharing of all values for long-term common interests. There is also, however, an active counter-trend of special demands for values for particular communities or organizations. Everywhere there remains the tendency to turn advantage into special interest for short-term benefit, a feature of the human condition that will always be with us. McDougal and his colleagues (1988, 972) conclude that the overall trend is "toward wider shaping and sharing of major values, despite zigzag patterns occurring at different times and in different communities. Yet the contemporary world has scarcely begun to mobilize its full potential to fulfill the rising common demands of humankind." Many value deprivations exist, and only a few communities enjoy a rich flow of values. The world social process, specifically the effective power process, is the condition within which this pattern plays itself out.

Conclusions

The social process is perhaps the single most overlooked dimension of policy making. Every detail is affected by interaction with the entire context, or social setting, of which the problem is a part. Therefore, part of your job as a professional, as a problem solver, must be to use a method that places the problem within a decision process within the setting of society. The inclusive method of social process mapping can be used to understand any conceivable interaction among people and to identify, describe, and account for the social details of any problem. Its categories are part of the policy sciences framework, which can guide you in examining and filtering information and thus inform your judgment. Without such a means to order information about who the relevant participants are, what they think, feel, and believe about the problem at hand, what outcomes they seek, and what institutions are at play, these social contexts can appear overwhelming. The basic idea of people striving to optimize value outcomes through institutions affecting resources is a particularly practical way for you to distinguish the power sector from all other value sectors, for instance, to identify value-institution outcomes, or to describe in functional terms the role of conventional structures and functions in any problematic situation. The real target, of course, is to solve problems and to secure common interests, but people's different ways of understanding problems and carrying out decision processes are direct reflections of the social process. The policy sciences call for explicitly, systematically, and empirically mapping this process.

SUGGESTIONS FOR FURTHER READING

Lasswell, H. D. 1971. Contextuality: Mapping the social and decision processes. Pp. 14–33 in *A pre-view to policy sciences*. New York: American Elsevier.

Lasswell, H. D., and M. S. McDougal. 1992. The social process as a whole. Pp. 335–73 in *Jurisprudence for a free society: Studies in law, science, and policy*. New Haven: New Haven Press.

McDougal, M. S., W. M . Reisman, and A. W. Willard. 1989. The world community: A planetary social process. *University of California Law Review* 21:808–971.

4 Decision Process
Clarifying and Securing the
Common Interest

All kinds of decisions are made daily about natural resources. Some result in immediate consumption of resources, whereas others defer utilization, some concern local areas, and some affect resource use globally. The outcomes and effects of our decision making will determine what kind of world we and our children will inhabit. Mapping the social process, the context in which natural resource problems occur, gives us the principal components of people's behavior—what makes up their differing perspectives, how they deploy their base values, how various situations affect their actions, and how to measure the outcomes and long-term effects of their actions in terms of what we call value-institutions. To close in on problem solving, though, we need to know how communities make decisions.

The second part of the conceptual framework of the policy sciences is the decision process, which identifies seven functions, or activities, of decision making. It is a continuous process involving many people, so we should not limit our attention to one decision or one actor. Careful delimitation of the decision-making events in any policy process can enable observers, analysts, or participants to understand what is going on and where improvements might be possible.

This chapter examines decision processes and their interconnected functions in terms of social process and functional value dynamics in an effort to improve understanding and influence decision making for better outcomes. It compares ordinary and constitutive decision making and suggests standards for judging their quality.

Overview of Decision-Making Processes

Natural resource policy and management is most usefully conceived as a process of decision making, and it is this process that must be upgraded to achieve better conservation and management. The *decision process* is a means of reconciling (or at least managing) conflict through politics in order to find a working specification of a community's common interests (Clark and Brunner 1996). Politics will always be with us because people pursue different policies reflecting their own standpoints, perspectives, and special interests. Nevertheless, people must and in many situations do reconcile their policy differences. Key outcomes of any decision process are the rules or norms it generates. The community to which they apply is expected to follow them. In fact, rules or norms are essential to coordinate community expectations and actions. Action is considered appropriate to the extent that it conforms with existing rules decided by the community based on members' shared expectations. Conversely, behavior is inappropriate when it does not meet the community's expectations.

Rules come in many forms for many different kinds of communities. They may be substantive or procedural. The rules for forest management in the United States differ from those in Indonesia, for instance, and the rules for acceptable professional behavior of range managers differ from those of medical doctors. Some rules may be informal, existing only in the form of social norms widely accepted by a group, whereas others, such as those established by experts for nonexperts (for example, requirements for graduation from a university or professional school), are more formal. Local, state, and national laws are highly formal rules. Still other rules are about making rules, such as the U.S. Constitution. The first three kinds of rules are produced by the ordinary policy process, but the last results from the constitutive policy process. Although the two blend together, the constitutive policy process is discussed separately later in this chapter. We are interested in both but will focus on ordinary processes because they are so pervasive in our professional lives.

The decision process can be conceived as producing rules or norms (Reisman 1981a, 1981b). Usually, certain activities or functions lead up to rule making and norm setting, and other activities or functions follow it. Thus, the decision process, which is continuous and nonlinear, can be seen with respect to particular decisions as a three-stage process of pre-decision, decision, and post-decision (Lasswell 1956, 1971a). In more formal terms, a decision produces a prescription, that is, a rule or norm supported by community consensus, which is to be enforced against challenges under various circumstances. The prescription is the way the community produces a policy that is acceptable for future implementation. An acceptable policy may be selected from among competing policies, it may be a compromise between competing policies, or it may be an integrative ("win-win") solution designed to defuse conflict.

The fact that a community produces a prescription does not mean that the common interest has been harmonized and that all special interests have gone away or ceased action. In fact, some community members may remain strongly set against the rules or policy. The term *consensus* does not mean agreement with the rules by everyone, only that all or nearly all expect the rules to be enforced by the community and its appointed officials. Consensus, especially on fundamental and complex matters such as justice, personal freedom, community laws, or natural resource management, is never total. Consider, for example, the history of finding and maintaining consensus on biodiversity policy—it has been a long and contentious process. To make matters more difficult, the detailed application of prescriptions may well vary from context to context.

If it is to be logically complete and explicit, a formal prescription should do five things (Brunner 1995a). First, it must be clear about the purposes or goals to be achieved. Second, it must specify the rules or norms of conduct intended to meet the purposes. Third, it should describe the contingencies or circumstances in which the rules or norms will be applied. Both the norms of conduct and the circumstances in which they will apply should be clear enough to be enforceable. Fourth, it should lay out sanctions to encourage people to comply with the prescription as well as penalties for noncompliance. And fifth, it should provide for assets or resources for administration and enforcement. If one or more of these conditions does not exist, the prescription will be ineffective in influencing people's behavior. For example, the 1989 draft management plan for koalas in Victoria, Australia, offered a long list of biological problems and alternatives and largely promoted further biological research (Clark, Mazur et al. 2000). Because its goals were not explicit, there was no basis for deciding which action to pursue and no way to assess the program's success. The plan did not adequately detail contingencies, sanctions, or assets for implementation. Furthermore, the plan had been in draft form for twelve years. It lacked force because it did not have an authoritative signature rendering it a final document, and government had not communicated its intent to implement the plan to the public. In short, this prescription was weak because it lacked some of the five elements of successful prescription.

The Seven Decision Functions

Like social process, decision process must be considered in terms of how it functions in order to develop a comprehensive and disciplined map (fig. 1.1; Burgess and Slonaker 1978). Regardless of the setting and its substantive details, every decision process invariably performs seven functions, whether they are properly carried out or not, whether participants are completely aware of them or not, whether decision makers are formally or informally organized, whether power in a community is broadly or narrowly held. Each decision function has characteristic, specialized institutions, and participants that carry out specific roles. Lasswell and Fox (1979) point out that in complex societies organizations that carry out each function are usually housed in buildings of characteristic architecture that symbolize that function. Because these are functions or activities of decision making and not stages, they are often carried out simultaneously, rather than sequentially, and they are often mixed together in complex ways.

This conception of decision making with its seven functions has undergone intensive and extensive testing over the past fifty years and has proved its utility many times. As Brewer and deLeon (1983, 21) note, the framework becomes more practical as its various interactive possibilities are explored, especially in actual cases. There have been other frameworks of policy making—the "rational comprehensive" and the "incremental, muddling through" frameworks, for instance (Lindblom 1959, 1980; Lindblom and Woodhouse 1993)—but the policy sciences decision framework does not lean toward either of these extremes. It accepts that in any policy-making process there will be elements of rationality and irrationality, comprehensiveness and parochialism. The decision process is essentially a political process with values at its core. Moreover, its analysis is in part a subjective matter. All participants, including professional natural resource specialists and analysts, should concern themselves with effective participation (Hogwood and Gunn 1986).

The following descriptions of the seven functions are based on Lasswell (1956, 1971a), Lasswell and Fox (1979), Reisman (1981b), and Lasswell and McDougal (1992). They are summarized in table 4.1.

Intelligence

Intelligence, often called planning, is the process of obtaining and processing information and giving it to decision makers and others. Decision making depends on information about past trends in events and the conditions under which those trends took place. Participants

Table 4.1. An Overview of the Seven Decision Functions, Standards, and Some Questions to Ask

Function	Standards	Questions to ask
Intelligence	Dependable (factual) Comprehensive (complete) Selective (targeted) Creative (in finding facts) Available (to everyone)	Is intelligence being collected for all relevant components of the problem and its context and from all affected people? To whom is intelligence communicated?
Promotion	Rational (standards) Integrative (synthetic) Comprehensive (holistic) Effective (adopted)	Which groups (official or unofficial) urge which courses of action? What values are promoted or dismissed by each alternative and what groups are served by each?
Prescription	Effective (expectations) Rational (balanced) Inclusive (includes all) Prospective (future-directed)	Will the new prescriptions harmonize with rules by which the community already operates, or will they conflict? What rules does the community set for itself? What prescriptions are binding? (These are easier to determine if they are written down.)
Invocation	Timely (prompt, open) Dependable (facts) Rational (common interest) Nonprovocative Effective (for application)	Is implementation consistent with prescription? Who should be held accountable to follow the rules? Who will enforce the rules? What sanctions will be applied in what situations? Are resources available to carry out the rules?
Application	Rational (meets rules) Union (contextual, unbiased) Effective (works in practice) Constructive (it helps)	Will people with authority and control resolve disputes? How do participants interact and affect one another as they resolve disputes?
Appraisal	Dependable (realistic) Continuing (ongoing) Independent (unbiased) Contextual (practical)	Who is served by the program and who is not? Is the program evaluated fully and regularly? Who is responsible and accountable for success or failure? Who appraises the appraisers' activities?
Termination	Timely (prompt) Comprehensive (holistic) Dependable (factual) Ameliorative (supportive)	Who should stop or change the rules? Who is served and who is harmed by ending a program?

Source: Adapted from Lasswell 1956, 1971a; Clark and Brunner 1996.

in intelligence must also project trends into the future (based on their understanding of future conditions) and invent and evaluate policy options to change conditions and, thereby, future outcomes. The goals sought by the community will determine what kinds of information about trends and conditions are relevant. Many sophisticated techniques exist to gather, process, and evaluate information, including field work, social surveys, models (both qualitative and quantitative), and pluralistic discussion. Your job as a professional is to make sure that reliable information is collected from all relevant sources and people and also that it is disseminated to all relevant people.

One case of an extensive and intensive intelligence activity that lasted for years was the reintroduction of wolves into Yellowstone National Park in the 1990s. Years' worth of detailed studies—largely biological but also sociological—were carried out. Qualitative and quantitative analyses were conducted. In some cases, computer modeling was done. Scores of public meetings were carried out in the region. Hundreds of articles were written. Slides shows and public information materials were prepared and distributed, and thousands of letters commenting on the details and wisdom of the proposed reintroduction were filed. All this was summarized in a series of formal government documents, including environmental impact statements that were distributed widely. These varied activities were all part of the intelligence and promotion functions associated with a decision process. Although it was protracted in this case, in some situations the intelligence function may be problematic because it is truncated or because information is not reliable or comprehensive. It is also not helpful if the information gathered is not relevant to decision making or if it is not made available to the right people. In this case, based on intelligence and associated open debate (promotion), a commitment (prescription) was made to reintroduce wolves, and invocation, application, and appraisals have since taken place or are under way. Termination has yet to occur.

Given the high level and intensity of interaction throughout society in general and in many natural resource fields, it is not surprising to find that planning is employed by every significant body of decision makers and that it is carried out with varying degrees of comprehensiveness. In many societies, the intelligence function is carried out by specialized institutions.

Promotion

Promotion is the function of recommending and mobilizing support for policy alternatives. Nothing is possible at the group or societal level without successful promotion. In order to meet the community's goals, people (sometimes the entire body politic) must be moved to act. Mass support and enthusiasm are essential in most cases to achieve collective action, whether at the local or the international scale. This may be true especially in times of emergency or when power is transferred from one holder to another (for example, at election time). Political parties, pressure groups, and powerful organizations are all specialized to promotion; much of the work of nongovernmental conservation organizations is focused on promotion, for example. When promotion is openly discouraged by authorities and when discontent is rampant, promotional activities may take the form of conspiracy.

Promotion should consist of open, active debate about what to do. When several different alternatives are advocated, proponents must mobilize resources, data, and public opinion to achieve their preferred outcomes. As promotion proceeds, people's expectations begin to crystallize and their demands are clarified. Among the questions you should ask about promotional activities are which groups (official or unofficial) are urging which courses of action, what values and what institutions are promoted or dismissed by each alternative course, and which groups are served by each alternative.

One example on a global scale has been the promotion in the past few decades of the environmental issue of sustainability. A landmark in this decision process was the World Commission on Environment and Development (WCED, otherwise known as the Brundtland Commission), which was created in 1983 to carry out intelligence, recommend solutions (promotion), and offer a prescription (see Dovers 1996 for a review). Eight principles were promoted to guide nations in their pursuit of sustainable development (WCED 1987): revive growth, change the quality of growth, conserve and enhance the resource base, ensure a sustainable level of population, reorient technology and manage risks, integrate environment and economics in decision making, reform international economic relations, and strengthen international cooperation.

The forum for putting the recommendations into action was the United Nations Conference on Environment and Development (UNCED, or "Earth Summit") held in Rio de Janeiro in 1992. This meeting, proposed by the WCED, was well attended by governmental and nongovernmental groups. It was the largest international gathering in history—more than eighteen thousand people participated, four hundred thousand visited the event, and eight thousand media representatives covered it. Agreements were reached among 179 countries.

The arena was an important open debate about the issue of sustainable development. The Rio outputs were voluminous and, although some prescriptions were authorized, the meeting was largely a promotional activity. Outputs included the Rio Declaration on Environment and Development (a list of twenty-seven principles), Agenda 21: Programme of Action for Sustainable Development (an action plan), the Statement of Forest Principles (a general statement of principles regarding forests), the Framework Convention on Climate Change (a document seeking to stabilize greenhouse gas concentrations), and the Convention on Biological Diversity (a statement on conservation of biodiversity). These are being promoted today in various ways, and some are being prescribed and applied at national levels.

Prescription

Prescription is the activity that establishes the rules by which people live. To prescribe is to clarify and articulate the basic goals and norms, or values, of the community. In prescription we formulate and enact policies or guidelines for action, crystallize the expectations and demands of community members, examine facts and their contexts, clarify rules, and examine the implications of rules. In turn, those with authority (the recognized right to act) and control (the means to encourage compliance with rules) must communicate and approve the rules. Decision makers must also determine whether new prescriptions will harmonize with existing rules or whether they will conflict.

Prescription comes about in any arena as people in a community come to hold common expectations about who is authorized to do what and how. These expectations are then enforced, sometimes severely, through sanctions. Prescriptions are made by various means in different societies. Even in folk societies without formal legislatures, this function still occurs. In more highly bureaucratized societies, people who carry out prescription include chief executive officers, advisors, councils, and legislatures, for example. Committee meetings in legislatures are often undertaken to inquire and discover which version of a contested prescription is to be accepted as true. Various people may be asked for advice—elders, scholars, and religious figures among them. Because modern life is complex, prescriptive bodies have invented many specialized organs; deliberations, committee hearings, and special investigations of specific subjects are common.

If your professional work contributes to this activity, be sure that the prescriptions you recommend contain the three key elements of content, authority signature, and control intent. Ask how these elements will be dealt with over the life of the prescription. Again, the content of a prescription must clearly state its goals (what is to be achieved), rules (what must be done to achieve the goals), contingencies (relevant situations in which it applies), sanctions (the indulgences and deprivations that will be applied for behavior that conforms or does not conform to the goals), and assets (resources that will be made available to achieve the goal and apply the rules and sanctions). In general, six kinds of sanctions can be identified (Lasswell 1971a): *deterrence* means that failure to comply with the prescription will result in specified value deprivations (for example, imprisonment), *restoration* seeks to bring back a situation that existed before deviations from the prescription occurred, *rehabilitation* helps people comply with the prescription (for example, counseling, job training), *prevention* encourages people to continue complying with the prescription (for example, rewards), *correction* seeks to change "deviational" individuals who are incapable or unwilling, despite educational opportunities, to conform to the prescription, and *reconstruction* seeks to rebuild public order.

The second necessary element of a prescription is its *authority signal*. In order to be taken seriously, a prescription must carry the stamp of authority, that is, those who establish the prescription must be recognized by the community to which it applies as having the right and power to do so. Third is the *control intent*. Authorities must communicate continuously their intent to apply the prescription and to impose sanctions if needed. Without these three elements, proposed rules will not become prescriptive.

In 1988, for example, the state government of Victoria, Australia, passed the Flora and Fauna Guarantee Act (FFG), a prescription to conserve the state's biological heritage. This act, which goes well beyond America's Endangered Species Act of 1973 as amended, is an innovative approach to biodiversity legislation, recovery actions, and community involvement in setting strategic direction for conservation. The FFG's content includes all five elements of good prescription— goals, rules, contingencies, sanctions, and assets—the goal being to guarantee that all taxa of Victoria's flora and fauna and ecological communities survive, "flourish and maintain their potential for evolutionary development in the wild." The act contains three major provisions: (1) to identify, conserve, and restore threatened and endangered species, (2) to identify and conserve imperiled ecological communities, and (3) to identify and eliminate "threatening processes." The FFG had the stamp of authority and a control intent when Parliament passed it, provided funds, and set up a new branch of government to put it into controlling effect. The new FFG Unit was then given responsibility for developing a practical strategy to invoke and apply the act, which it has done.

Invocation

Invocation is the first action taken to invoke, or appeal to, a prescription. A community's efforts to put a prescription into effect, its determination that a prescription is consistent with established rules, and its decisions about who should be held accountable to follow the rules and who will enforce them are all part of invocation. More technically, it is the initial or provisional characterization of the kinds of behavior that will be permitted in terms of a prescription. Invocation sets up administrative arrangements; it allocates people, resources, and facilities. Individuals or groups may disagree on how to invoke a particular prescription, and their actions may be subject to reviews by higher authority, which moves into the application function. There might be disagreement, too, about how well invocation is proceeding, and there may be complaints to officials and a decision to take a case before a full-fledged administrative hearing or commission.

Societies have people, buildings, and institutions specialized to this function as they do for all the other decision functions. One obvious specialization involves security and enforcement activities, including the police. Highway patrol officers, for instance, enforce speed laws and other prescriptions that set out rules for operating motor vehicles on roads. Generally, people in these positions are authorized to decide provisionally which acts are violations of a prescription and to take violators into custody. For example, when a game warden cites or arrests a fisherman for taking more than the legal limit of fish, we acknowledge that the warden has the legal authority to make the arrest, but the judgment that the fisherman violated the law is, at that point, only tentative or conditional. The warden does not convict or sentence the fisherman; those final, more conclusive actions are part of the application function. People who carry out invocation are also usually empowered to hear and to act on complaints of violations, and, being aware of the threat of violence and crime in society, they are skilled in coping with disturbances coercively. In repressive societies, police (which may be hard to separate from the military) may assess individuals' loyalty to the official elite or to the state.

Whatever the arrangements for invocation, you should make sure that they constitute effective, practical, and complete ways of executing the prescription. Do the arrangements make adequately clear exactly what the invocation activities will be in different contexts or under different contingencies and who will be held accountable to follow the rules? In other words, who does it and to whom will it be done? Are the people who carry out invocation adequately skilled, and do they receive ongoing training? Are the organizations adequately staffed, and do they have sufficient resources? To whom should they turn (other than those doing application) if they run into problems in carrying out their invocation activities? How can the invocation function be appraised as a basis for improving it?

Returning to our example from above, the Flora and Fauna Guarantee Act established practices for its own invocation. The act provided an administrative and management framework for conserving all taxonomic groupings of Victoria's flora, fauna, and ecological communities. All state government and public authorities are bound by the objectives of the act, which include a wide range of mechanisms to facilitate implementation. One of the first tasks of the newly established FFG Unit was to draft a Flora and Fauna Strategy that gave a statewide perspective on issues, priorities, and directions for achieving biodiversity conservation. Other invocation activities included preparation of detailed management plans for listed items, mechanisms to protect critical habitats, the requirement for community involvement, and controls over the taking, trading, and keeping of native flora and threatened fish species. A listing process was set up for legal recognition of threatened biota and potentially threatening processes. An independent Scientific Advisory Committee composed of government and nongovernment members was set up to consider all nominations for listing. Any person can nominate an item for listing, and all proposed listings are published for public comment. The minister makes final decisions. A similar delisting process has been established.

Application

Application is the final characterization of people's behavior in terms of a prescription in specific situations. Application ideally resolves disputing claims over how prescriptions will be implemented and who has authority to make those decisions. As a result, it covers final decisions, such as those made by formal entities (for example, supreme courts, high-level regulatory commissions, or chief executive officers). There may be no sharp line dividing invocation from application, although the later stages of task completion unambiguously involve application. Claims about departures from prescriptions—based on peer review, authority, or other mechanisms—are resolved in application. Participants must interpret the rules in particular cases, supplement them if needed, and integrate old and new prescriptions into a working program. There must be enforcement as well as continuous review and approval or disapproval of behavior at individual and organizational levels; these are the tasks of application and invocation. This function should be carried out in open, pluralistic arenas. The courts may figure prominently, but many resolutions take place formally or informally within the daily operations of programs. Application involves decisions about when disputes will be resolved by people with authority and control and how participants should interact and affect one another as they resolve disputes.

The application function is obvious in many natural resource policy issues in the United States. Many complex, diverse, and highly contentious issues—including endangered species, water management, mining rights, and management of parks, forests, and grazing lands—have ended up as high-profile, closely watched cases in which the courts have attempted to interpret prescriptions, resolve differences, and apply sanctions. A number of these disputes have centered on questions of federal versus state control of natural resources and establishment of authority relations.

Most application activities are carried out formally by specially designated people (magistrates, judges, high-level officials). They typically follow precedents and established guidelines in making judgments about whether invocation was carried out well. Many disputes about how to invoke and apply prescriptions are more informal. If you are involved formally or informally in application activities, you should ensure that application is done constructively so that it actually controls people's practices and puts the prescription into effect. Ask whether it contributes to better social and decision processes in the future and whether it mobilizes consensus. Is it fair? Is it consistent with the community's common interests and free of biases?

Continuing our Australian example, the application of the FFG flows from earlier prescriptions and invocations of actions. For example, Action Statements (recovery plans, or prescriptions, of six to twelve pages) document high-priority actions needed to conserve threatened taxa and communities or to mitigate the effects of threatening processes. Part of this effort is to identify and address social and economic issues in recovery planning, which means achieving conservation objectives with minimal negative impact, clarifying and ranking alternative strategies, and evaluating the impacts and costs of particular actions (Crosthwaite 1995). Conservation is achieved through various invocation activities (policing, enforcement, on-ground work, and community education) in specific contexts (see Mansergh et al. 1995).

But any new legislation takes time to achieve broad acceptance before it becomes routine. Businesses, local governments, and conservation groups have criticized the Action Statement process, focusing on lack of public input, competency of biological assessments, and adequacy of socioeconomic input. Although some of these concerns were the result of misunderstandings of the role of the Action Statements and how to apply them in the overall process, the criticisms were used to modify the process and significantly resolve disputes—part of the ongoing application function. Application and invocation of the FFG were appraised by Wilson and Clark (1995), who concluded that these two functions could be improved by development of critical habitat determinations, wide use of techniques such as population viability analysis, provision of adequate resources, and regular overall appraisal of the act's implementation.

Appraisal

Appraisal is the assessment of a decision process as a whole and of the success of particular prescriptions in achieving their goals. Through appraisal, communities can estimate the degree to which their public policy goals have been reached, assess the causal factors involved, determine responsibility and accountability for what happened in a particular decision process, and ensure that their findings and recommendations are distributed appropriately. Because appraisal reflects on the judgment and competence of public officials who have been responsible throughout the decision process, these same officials are often highly reluctant to delegate this task to anyone else. As a result, appraisals may be rare and limited in scope—so much so that formal censors are sometimes set up to investigate the performance of officials and even to undertake actions that may lead to sanctions. Nevertheless, appraisals are under way informally or formally in all societies all the time.

The basic criterion is the policy objective that was originally sought. This objective tells appraisers what they should consider in assessing the situation. Means vary among societies. In folk societies, elders and their recollections are the chief repository of goals and instrumental prescriptions, but in modern societies, policy goals are usually written and thus available to wide audiences. From these sources various means of appraisal are available. It is possible only if aggregate information is gathered and processed. Big policy requires big appraisal, which means large volumes of pertinent facts and figures. Technology, especially the computer, is used in many evaluations of decision process. Appraisals help every community update its self-image in detailed and comprehensive ways. Trend data are basic to appraisal, as they are to intelligence; whereas the former deals analytically with the past, the latter looks to the future as well. In appraisal, the task is to provide a dependable basis for inference about formal and causal factors involved in how a policy was carried out or applied. Because of their importance, appraisals should be independent and ongoing, not limited to the end of the process.

In natural resource policy and management, appraisals are an important means of assessing whether a prescription has effectively met its goals and who is responsible and accountable (see Clark 1996a; Winks 1997; Margoluis and Salafsky 1998). They represent a major opportunity for learning and course correction, for finding the lessons of experience and translating them into the lessons of foresight. For instance, with the continuing growth in the use of community-based initiatives, it is vital to improve their effectiveness by appraising why some CBIs work well and others do not. Robar (1999), Berry (2000), and Wondolleck and Yaffe (2000) are among those who have appraised this general approach, and Cromley (2001) has evaluated the utility of CBIs in a specific case, namely, bison management in the Greater Yellowstone ecosystem.

Another example is the high-profile Leopold Report of 1963, which was a review of resource management by the National Park Service. Because it was conducted by outside experts, it had a greater impact than the numerous reports before and since on this matter from within the Park Service. The Leopold Report received widespread publicity and was reprinted in many national publications, including the Sierra Club Bulletin. The review resulted in a new set of principles (prescriptions) that essentially called for each national park to become an "illusion of primitive America" and for the Park Service to create or preserve a "mood of wild America" (Sellars 1997, 214). The Leopold Committee's appraisal placed management of America's national parks on new philosophical grounds, involving broad ecological principles for the first time. The Park Service took the Leopold Report as a major challenge, a call to arms, so to speak, and since adopting its principles has faithfully tried to apply them.

Termination

Termination is the repeal or large-scale adjustment of a prescription. It involves canceling or succeeding the original prescription (and modifications made in the course of its invocation and application) and the social structures set up for its implementation. It may also involve compensation to those affected by the changes. Termination ends prescriptions, practices, or policies that have accomplished their goals as well as those that are not working and clears the way for a new beginning. This function is often overlooked or underappreciated in many policy processes, and despite its overwhelming importance, there are very few structures specialized to this function.

In all cases, attention must be given to the people who will be affected by termination. As professional participants in a decision process, you need to consider who will be served and who will be harmed by ending a program, and you must deal with people's fears and hopes (expectations of indulgence or deprivation). You should also determine what needs to be done so that termination can be carried out smoothly. Difficulties that arise from misunderstanding, deceit, or intimidation must be avoided if termination is to be kept fair and timely. When government is involved in termination—think of major changes in federal regulatory practices or the cancellation of large entitlement programs—negative sentiments about the termination of a program or policy are often transferred to the government. But governments are generally becoming more concerned with the interests of their people because of a growing awareness of the hardships sometimes caused by changes in public policy.

One example of failure to attend to termination activities is found in endangered species conservation. The Endangered Species Act sets out methods to start up recovery programs, and the federal and state agencies and courts are experienced in the invocation and application functions, but little attention has gone into planning or implementing successful terminations. Although very few species have been removed from the protection of the Endangered Species Act because they have been fully restored to viability in secure habitat, the "delisting" process and post-recovery plans, programs, and actions need considerably more systematic, comprehensive treatment than they have received to date. In the American West, efforts to delist grizzly bears since the late 1990s have been contentious and protracted; the situation for wolves will likely be the same, although some participants have begun to plan for that eventuality.

Ordinary and Constitutive Decision Making

In understanding how decisions are made, it is helpful to distinguish between ordinary and constitutive decision making. Ordinary decision making focuses on substantive policy choices, and deliberation takes place within a given structure or context of decision making that is viewed as legitimate. In contrast, constitutive decision making is made up of deliberations and choices about how policy should be made and, by implication, who ought to be involved in the decision process. Thus constitutive decision making goes well beyond the everyday operation of existing decision processes to look at how institutions, analytic techniques, procedures, and participants ought to be structured or selected (Lasswell 1971a, 77, 98–111).

Robert Healy and William Ascher (1995) make this distinction in an interesting study on the impact of increased information in U.S. Forest Service planning. They observe that over the course of the previous two decades there were significant additions to knowledge about natural resources, but that new information did not always lead to improved policy via ordinary processes and was often not an eagerly awaited source of consensus. Despite the basic value of reliable intelligence, it is often used selectively to promote special interests. In forestry, information has also become the raw material for partisan use by a host of self-interested participants. Finding ways to obtain and use information to clarify and secure common interests remains a challenge in ordinary decision process. Healy and Ascher conclude that this process is affected in several distinct ways by changes in the availability of information about natural resources, that "although new information may change policy outcomes, often for the better, there is little reason for believing that it will make the decisionmaking process itself either shorter or smoother" (17), and that if improvements are to be made in knowledge use, changes in constitutive as well as ordinary processes may be needed.

The Ordinary Decision Process and Standards for Decision Making

The object of learning about decision process is for you, as a natural resource professional, to bring about improvements in the formulation and execution of policy. There are many examples of professionals grappling with improving decision process for parks (Brandon et al. 1998), public health (Raffensperger and Tickner 1999), and wildlife habitat (Noss et al. 1997). Each of these policy arenas and many others function more or less adequately. But to be most effective, Harold Lasswell argued, in today's increasingly complex world it is vital for responsible decision makers—and by this he meant professionals as well as other participants—to subject themselves to the discipline necessary to stay in touch with reality and to appreciate that there are always several versions of reality in competition in policy making. This is one way to keep decision processes effective. Decision makers and other participants must find ways to be realistic and creative if they are to get beyond the reams of computer printouts and summaries by diverse experts and special interests. In short, you need a comprehensive conceptual map to guide your attention to the important matters. Such a map, which draws on the decision functions listed above, also helps you to focus on standards of performance. Having explicit criteria can expose mistakes, inconsistencies, contradictions, and deceptions in decision processes. It can also make clear how each decision function might be improved, according to specific standards, in practice.

Some of the criteria that Lasswell (1971a, 86–97) recommends pertain to particular decision functions in both ordinary and constitutive processes, whereas others are broadly applicable to all seven (see table 4.1). These standards can guide your appraisal and actions to improve natural resource policy and management.

Dependability: Statements of fact made available to people in the decision process must be reliable and realistic. If there is doubt, then a panel or committee should assess the credibility and qualifications of the experts making the statements.

Comprehensiveness: The decision-making process must attend to all data about the problem, including historic trends, conditions and causes, and projections about the future. The results of this research must be inclusive (that is, including all people) in terms of the goals sought and the analysis of alternatives.

Selectivity: The results of the analysis should be targeted to the problem itself and to people affected by it.

Creativity: It is necessary that new and practical goals and strategies be carefully compared with earlier or less realistic alternatives. Both experts and citizens can make these determinations.

Openness: The decision process, especially the intelligence function, must be kept open and cooperative. Efforts need to be made to include the public and other specific individuals and groups, both as sources of intelligence and as recipients.

Rationality: This is a major issue related to our emphasis on clarifying and securing the common interest. Proposals and justifications must be rationally complete and analytically sound so that judgments can be made about them. It is important to know what information was used, whether anything was overlooked, and whose scope values were served.

Integrativeness: It is highly desirable to have an integrated approach that the people involved do not view as a patchwork of compromises. Integrated solutions mean "win-win" approaches.

Stability of expectation: The continuity of public order must be sought. The decision process should support people's expectation of order and strengthen their confidence in public order.

Timeliness: Action must be taken in a timely fashion, especially when individuals or groups are seriously deprived. Participants in the decision process should deal with justifiable complaints promptly, and they should map trends in complaints, especially from people who are not fully enfranchised in the process. One key indicator of how well the overall decision process is working is whether there are delays between when a problem is raised and when it is addressed. Delays are evidence of a poorly performing process.

Nonprovocativeness: The outputs, or results, of invocation and application should impose no additional deprivations than are minimally required.

Realism: The decision process should seek the common interest as formulated in the rules prescribed above. The steps and parties involved should be reviewed to make sure there is conformity to the rules.

Uniformity: The process must be applied without discrimination. Participants should make sure that there are no controversies about unfairness.

Independence: Appraisal or evaluation must be free of pressure, threats, and inducements that might bias the outcomes. Both in-house and independent appraisers are needed.

Continuity: Appraisal can be done on an occasional basis, but it is most helpful when it is continuous.

Balance: Because the decision process usually results in change, participants should try to balance necessary changes with disruptions of existing operations.

Ameliorativeness: Deprivations and coercion should be avoided. Participants should consider whether they could facilitate the process by compensating people who have been deprived by changing a program.

Lasswell (1971a) recommends several additional criteria that apply to all functions and structures.

Economy and technical efficiency: The decision process should be carried out in a cost-efficient way that meets technical standards.

Honesty and reputation for honesty: People involved in the process should be honest and be recognized throughout the community for their honesty.

Loyalty and skill of official personnel: One goal is to inspire and deserve confidence in the integrity of the decision-making process, and this, of course, requires high-quality people.

Complementarity and effectiveness of impact: This means that every decision function should be carried out in such a way as to foster continuing support for the political system as a whole and to realize the common interest of the whole community.

Differentiated units within an organization: It is important to have separate organizational units for each of the seven functions, but they should be configured in a coordinated and integrated manner. In large organizations in particular, a unit for each decision function is likely required to ensure the independence and vigor of the operations involved.

Flexibility and realism: The process should be flexible and realistic in meeting goals in times of crisis as well as during intercrisis periods.

Deliberateness and responsibility: The whole decision process should be conscious, intentional, and considered, not erratic or impulsive.

The Constitutive Decision Process and Five More Standards

The constitutive decision process is about deciding how to decide. It lays out for any community how the common interest will be clarified and secured (McDougal and Reisman 1981). Concerned with the establishment of "authoritative and controlling expectation about who, acting how, is justified in making enforceable decisions" (Lasswell 1971a, 99), it becomes the means to resolve uncertainties and conflicts in all subsequent decision making and thus is an especially important issue under any circumstances.

The overriding constitutive document in the United States is the Constitution, which lays out a framework for the ordinary decision processes that have taken place daily at all levels of our self-governance for more than two hundred years. But other constitutive processes occur, nearly all, by definition, on a smaller scale than the making of the Constitution. For example, every time an endangered species recovery program or any new management and policy effort is established, the participants must decide how they will make all subsequent decisions. This is a constitutive process. It may be lengthy and deliberative or it may simply borrow an "off-the-shelf" approach used elsewhere.

Constitutive decision making has criteria for performance that are the same as those for ordinary decision process, plus five additional ones having to do with the preferred allocation of authority and control (Lasswell 1971a, 105–11). First, the constitutive process should arrange for common interests to prevail over special interests. This is the central and perennial concern in the allocation of power and control in all systems of public order. The goal is to discover and work toward common interests, and the procedural solution is to be inclusive and involve all interested people all the time. This is democracy. Knowledge has to be mobilized, which requires support for libraries, community planning exercises, laboratories, and field experiments. It also requires notifying the public, holding public hearings and inquiries, and having consultative and advisory services broadly available to the public and to decision makers.

Second, the process should give precedence to high-priority rather than low-priority common interests. Decision makers should be aware of which interests are more widely shared than others. The public is often in general agreement on certain matters. For example, meeting the common interest prescriptions of a sustainable global environment as detailed in the Rio principles should take precedence over less inclusive common interests.

Third, the process should protect both inclusive and exclusive common interests. Common interests are inclusive if they genuinely and significantly affect all people. If they predominantly affect one or a few participants or only part of a community, they are exclusive. Both must be protected. Inclusive common interests should not preclude exclusive interests and vice versa. "The common interest is served by a public order that protects freedom of choice and leaves as many decisions as possible to parallel actions outside the realm of direct collective action" (Lasswell 1971a:108).

Fourth, when there are conflicting exclusive common interests, the constitutive decision process should give preference to the participants whose value position is most substantially involved. When individuals or subgroups are unable to resolve their conflicts on their own and there is increasing potential for damage to the public order, the larger community must become involved. The courts, for example, may serve to resolve such conflicts and at the same time support the public order. Government regulatory functions serve the same purpose, although, unfortunately, regulatory functions are often perverted to serve special rather than common interests (whether inclusive or exclusive). Many examples of this practice litter natural resource management arenas. This principle supports the policy sciences' preference to give priority to individual and subgroup choice wherever possible.

Fifth, in addition to authority, the process should allocate base values of sufficient magnitude (power, money, knowledge, skill, and so on) to enable authority to be controlling. This means simply that authoritative bodies should be given ample power and resources to do their work. Money alone is not enough; it must also include an authority signature and control intent. Power allocations must be proportional to the task at hand, and the public must participate in the overall authority process in inclusive ways.

In summary, the constitutive decision process is of fundamental importance because it shapes how all ordinary decision processes will be carried out. As Lasswell (1971a, 110) notes, the problem of the design of constitutive power structures is to find relatively coherent and independent activities that continuously provide focus for public attention and a clear set of common values and expectations. The ordinary and constitutive decision processes are the means society uses to distinguish common interests from special interests and to achieve the former.

Mapping Decision Process

The reason for learning about decision processes is that often there are malfunctions in one or another of the seven functions, and part of your job as a professional is to identify these and figure out where, when, and how to intervene to improve a process and its outcomes. Remember that the decision process is the very heart of policy making. Social process is the context within which it takes place, but decision making itself is key. It is a way of clarifying and securing common interests, despite diverse pluralistic interests, special interests, bounded rationality, incomplete and distorted information, internal and external surprises, and other factors.

A Method for Mapping Decision Process

The following suggestions for mapping decision process are based on an outline by Clark and Willard (2000). This generalized, easy approach can be a guide to studying and writing about any decision process (see table 4.1).

To begin mapping your case, familiarize yourself with its details, both procedural and substantive. You need to ask where in the decision process a breakdown has occurred. What functions have not been carried out well? Why are the community's goals not being met? You need to identify pitfalls in the decision process, formal or informal, that are responsible for problems. This is the point at which you actually "map" the decision process. Look closely at your case and catalogue what activities have been carried out in order to perform each function. What intelligence activities have been undertaken? What promotional activities are under way? Has a decision been made in the form of a prescription? What prescriptions already apply? Go through all seven functions, noting outcomes of each decision function and how they were reached.

By using the performance criteria described above, you should be able to pinpoint where certain decision functions have not been well executed and to determine what factors are formally or effectively responsible for any major malfunctions. One way to think about this is to consider the seven categories of social process—participants, perspectives, situations, base values, strategies, outcomes, effects—and to focus on changes in value-institutions within the decision process. How are the eight values being shaped and shared through the various decision functions? Is decision making in your particular case carried out in favor of "a commonwealth of human dignity," or is it moving away from it? Just a few of the questions you should ask about the intelligence function: Which data are being collected and which are being discounted? Whose perspectives are being taken into account and whose are being ignored? Who has the power to authorize intelligence gathering? Who has the skills to collect reliable intelligence, and who has the knowledge to interpret it? Does the information have power and wealth (or any other value) implications? Who gets access to this information? How is the situation or arena organized to gather, process, and disseminate intelligence? Which strategies are used in these three activities and by whom? Looking at the outcomes of the intelligence function, whose value positions are being indulged and whose are deprived? In terms of what values? Are there likely to be any permanent institutional changes as a result of how this intelligence activity was carried out? Ask these kinds of questions about each decision function.

Finally, you must devise a strategy to solve the problems. Your recommendation will constitute a "case-specific theory" about how to improve or resolve matters. You need to figure out who should intervene in the decision process, when, where, how, and what should be done to improve policy outcomes. You should be able to justify your proposed solution by explaining why it is likely to work. You should specify what resources would be required. Remember that it is important to review the decision process as a whole as well as distinguish the seven functions; a breakdown in appraisal, for example, may be caused by a weak intelligence function.

For example, I was part of a team of independent analysts in a project to develop better policy for koala conservation and management in Australia (Clark, Mazur, Cork, and Harding 2000; Clark, Mazur, Begg, and Cork 2000). We began by asking questions—to fulfill the intelligence function—such as, What do we reliably know about koalas, their biology, populations, and habitat requirements? What do we reliably know about how the public values koalas? How do commercial and other human activities affect them? How well is existing management policy conserving koalas? How is information about these and related topics gathered, processed, and disseminated? If the information on these topics is adequate, what should be done, and if not, then what? In short, we examined all the categories of social process—participants, perspectives, situations, base values, strategies, outcomes, and effects—as they pertained to each of the decision functions.

Questions we asked about the promotion function included the following: What recommendations are being made by the Australia Koala Foundation, by the federal government, by the states, by the scientific community, and by other groups? Which of these will most likely address the problems that threaten koalas? What values are being advanced or moved back, and who is being served by each of these various proposals? What strategies are being employed? Again, we looked at all relevant categories of social process and carried out a thorough problem orientation. The overriding question was how the common interest might be advanced. Similarly, we asked case-specific questions about the prescription, invocation, application, appraisal, and termination functions. Throughout the project, we used this systematic approach and its recognized standards to appraise the overall decision process. This analysis, based on a considerable amount of empirical research, and our recommendations for improving koala conservation are but one case in which the policy sciences framework has been employed to understand decision process and find practical ways to serve the common interest.

One way to learn how to map decision process is to learn how other people have done it. Ten cases in Clark, Willard, and Cromley (2000) offer diverse examples from around the world, two of which are summarized below to illustrate use of the categories. The first looks at decision making in two ecotourism projects, one in Central America and the other in Africa, and the second illustrates how weaknesses in the process have limited solutions to the problem of urban ozone pollution in Baltimore, Maryland. Clark and Willard (2000) further describe decision process, and Clark, Casey, and Halverson (2000) discuss the decision process for elk management on the National Elk Refuge in Wyoming. There are many other helpful examples of people carrying out appraisal and offering recommendations to improve policy (for example, Yaffee 1982, 1994a, 1994b; Alvarez 1993, 1994; Miller et al. 1996).

Ecotourism in Central America and Africa

In recent decades, ecotourism, a type of specialty travel, has become increasingly popular and has been justified as a tool for protecting biodiversity. Income from tourists for entrance fees, guiding services, hospitality services, and sale of souvenirs is expected to be used to protect natural wonders and to compensate local people whose livelihood previously depended on using or transforming local resources. The underlying premise of so-called Integrated Conservation-Development Projects is that local people will support conservation efforts if they are involved in their development and will benefit from them. In practice, however, no universal guidelines for responsible ecotourism exist, and it looks as if many ecotourism programs are failing to achieve the goals of biodiversity conservation and local participation. Eva Garen (2000) compares the decision process in an ecotourism project on the island of Roátan, Honduras, with the Mountain Gorilla Project in Rwanda to learn why the highly touted ecotourism industry has been less than successful.

According to Garen, tourism on Roátan's coral reefs expanded rapidly without prior planning. The intelligence function was dominated by local elites interested primarily in wealth and power, and little effort was made to obtain social or ecological information prior to the influx of tourists. Promotion was similarly monopolized by a small group of participants with short-term financial interests. Effective guidelines for protecting biodiversity were either nonexistent or poorly implemented. Garen notes that

> prescriptions were established through a series of open and inclusive town meetings to limit direct tourist impacts, to determine where tourists could snorkel and deep-sea dive, and to minimize shoreline development. The community, however, did not have adequate funding, so their efforts were dependent entirely on funding from the local elite. Although these elite groups appeared to support conservation efforts, they did not give reserve staff the authority to invoke protective prescriptions. For example, although local residents were hired as patrol officers, they rarely were taken seriously in their attempts to make provisional citations or assertions of control. (231)

Termination of established practices (economic, symbolic, or otherwise), power structures, or program prescriptions is seldom considered in ecotourism programs, nor is appraisal systematic, unbiased, continuous, or fully contextual. In the case of the marine reserve at Roátan, the local elites and foreign tour operators conducted their own assessment, which concluded that the reserve had been improperly run and that they would no longer fund it—appraisal and termination activities that clearly served their own purposes.

The Mountain Gorilla Project in Rwanda is one of the best known ecotourism programs. Garen feels that, although reputedly successful, it cannot be held up as a prototype of decision making because it was not clear from the published literature how the decision process was conducted. Nevertheless, the case had a number of positive elements. The project began in 1978 when the New York Zoological Society sent two zoologists to work with the Rwandan government to determine why the gorilla population was declining and to recommend corrective action. Their goals were to study the land use practices of gorillas and humans, identify attitudes and values of local people toward the gorillas and the park where they lived, develop a tourist program focused on the gorillas, and initiate a conservation education program.

The researchers compiled an impressive amount of background information. They also realized that a successful ecotourism program must ensure that the gorillas would not be harmed by tourism, that adequate revenue would be generated, that local people would benefit, and that the Rwandan government would place a high priority on mountain gorilla protection. But Garen points out that it is not evident how these decisions were made, that is, who makes natural resource decisions in Rwanda, who was involved in analyzing and interpreting results of the social and ecological surveys, how local people were involved with program development, or how and with whom the researchers clarified goals, described and projected trends, analyzed conditions, and formulated policy alternatives. The point is that, although protecting native species and helping local people were the goals of the researchers and the New York Zoological Society, there is little evidence that local people were actually involved in the decision making. In other words, it is not clear from the written record that the process by which these goals were decided was inclusive, open, and democratic, or that it served the common interest of the people most affected by the proposed changes.

The several prescriptions that were made seem to have been strong in terms of content, authority signal, and control intent, and invocation and application seem to have been largely successful. For instance, poaching and pressures to clear land were alleviated by border patrols, tourist contact with the mountain gorillas was strictly controlled, substantial entrance fees were collected (although the government rejected the recommendation that it share revenues with the local communities), and an education program was initiated (although the government halted student visits to the gorillas because that practice limited its revenue from foreign tourists). But again, the question arises of who participated in deciding on these prescriptions. Moreover, Garen was unable to determine from the literature whether termination of local practices and power structures was carried out in timely, dependable, comprehensive, or ameliorative ways. Nonetheless, appraisal by the program developers a few years later showed great gains in meeting the program's goals of increasing numbers and survival rates of gorillas, significant income for the government and local communities, major growth in high-paying, full-time jobs, greater recognition of nonconsumptive uses of the park, the forest, and the gorillas, and government support for the program.

Garen notes several lessons for the ecotourism industry from these two case studies. It is important to examine the actual processes through which decisions about natural resources are made in specific contexts and to document these processes. "Successful" outcomes do not necessarily indicate good processes and may reflect the values of those doing the appraisals more than those of the communities involved in or affected by the programs. Social process dimensions of the industry must also be understood—people's perspectives, values, and interactions with resources as well as their power struggles. Finally, it is important for the industry and the public to examine the appropriateness of ecotourism as an integrative strategy for conservation. She recommends systematic appraisals of decision processes in ecotourism programs in many different contexts from which truly sustainable ecotourism prototypes can be developed.

Ozone Pollution in Baltimore

The second case focuses on Baltimore, where residents suffer significant morbidity from exposure to high levels of ground-level ozone pollution. Alejandro Flores (2000) looks at key points in the policy process where improvements could be made to reduce health risks. He identifies a multitude of organizational participants, noting that official and conspicuous actors may garner attention at the expense of other "unofficial, inconspicuous or unconscious" participants who also have an impact on value outcomes. Among government participants are agencies dealing with environmental protection, health, transportation, planning, natural resources, and housing. In addition there are point-source and non-point-source polluters, hospitals, media broadcasters, health associations, environmental groups, and universities. After detailing the evolution of the decision process, specifically value

outcomes and prescriptions, since the discovery of the pollution problem in the 1950s, Flores examines factors that have shaped trends in
outcomes.

The formation and accumulation of ozone, human exposure to it,
and, ultimately, its health impacts are very complex processes that involve a number of social and ecological facts that are not part of the
ozone-abatement decision process alone. For instance, the formation
of ozone and emissions of its precursor pollutants have been decreased
by gains in fuel and combustion efficiency and offset by increases in
the numbers of automobiles and miles driven. Accumulation of ozone
is checked by "sinks" (forces that limit its buildup), including wind,
which carries the pollutants away, and soils and vegetation, which "sequester" the ozone. These have also seen changes over the past forty
years. Although removal of ozone by wind currents is not under the
control of any decision process, many changes have taken place in vegetative cover and soils in urban, suburban, rural, and upwind areas of
Baltimore. Finally, changes have occurred in the human health impacts of ozone that have not been a result of the abatement program
(and its public education program). Demographic changes include suburban increases in population, urban decreases, and changes in age
groups, as well as changes in amounts of time that residents spend
outdoors (even as related to changing weather patterns).

Flores finds a number of weaknesses in the decision process for
ozone abatement. Intelligence activities are narrow in scope, largely
ignoring measurement of human exposure and spatial and temporal
heterogeneity in all phases of formation, accumulation, and exposure.
Promotion is hampered as a result, especially since most promotion
takes place elsewhere (in Washington, D.C., where the federal government makes abatement regulations). One of the problems in prescription is the use of narrow, uniform rules established for the entire country by the Clean Air Act and directly adopted by the state of Maryland.
The biggest problem in prescription, however, affecting the entire policy process, has been "goal substitution." Because health problems
were originally attributed to emissions, the goal was framed not as the
protection of human health but as the control of emissions, and as a
result, alternative solutions that addressed additional causal factors
were neglected. The consequence has been adoption of national air
quality standards that are insensitive to local contexts and seriously
restrictive in potential courses of action (Flores 2000). Invocation and
application have shown repeated delays—in deadlines for meeting
standards, in policing, and in applying sanctions. Moreover, the actual
victims of ozone-related problems have no say in implementation of

abatement programs. The author also finds considerable deficiencies in appraisal: it is neither independent nor comprehensive in terms of the phases of ozone dynamics, the multiplicity of participants, or the full range of decision activities.

Finally, Flores's study suggests three policy alternatives and key actions for implementation. The first is to improve ongoing efforts to control emissions, but this is unlikely to achieve the ultimate goal of improved health because federal standards have recently become stricter and emission reduction is becoming increasingly expensive and difficult. The second is to introduce measures to offset losses in rates of ozone removal—in other words, increasing vegetative cover in and around the city, which will take up and sequester ozone from the atmosphere. The third suggestion is to minimize human exposure, particularly of vulnerable individuals in highly affected areas during periods of highest ozone levels. Fundamentally, the problem that exists in human health in the Baltimore area occurs because of undesirable outcomes of past decision processes. If policy goals were redefined to reflect more accurately the goal of improving human health, the range of alternative solutions would thus be enlarged.

Conclusions

The decision process framework is invaluable for permitting professionals to distinguish functions of decision making, identify who is performing them, and judge whether they are being performed well. You will be able to see, for example, when environmentalists, industry representatives, and government agents keep advancing their own solutions without listening to each other, that the process is stuck in promotion. Or, when a state agency, mired in public censure over an issue, hires a former employee to evaluate its performance, that the appraisal is neither independent or ongoing. In addition to examining each function, you will be able to step back, take a longer view, and look at the overall decision process. Your findings will put you in a strong position to recommend where and how the various decision functions or the overall process can be improved. Knowledge about the standards of rule making can contribute to more successful decision processes and, as a result, broader public support of and trust in the process. As Lasswell (1971a) notes, rational decision making relies on a clear conception of goals, reliable calculation of probabilities, and skilled application of knowledge of ways and means. The decision process framework encourages professionals to enhance the rationality of problem solving in the common interest.

SUGGESTIONS FOR FURTHER READING

Clark, T. W., N. Mazur, S. J. Cork, S. Dovers, and R. Harding. 2000. Koala conservation policy process: Appraisal and recommendations. *Conservation Biology* 14: 681–90.

Cromley, C. M. 2002. Beyond boundaries: Learning from bison management in Greater Yellowstone. Ph.D. diss., School of Forestry and Environmental Studies, Yale University.

Lasswell, H. D. 1971. Professional services: The ordinary policy process, pp. 76–97, and The constitutive policy process, pp. 98–111, in *A Pre-view of policy sciences*. New York: American Elsevier.

5 Problem Orientation
Focusing on Problems to Find Solutions

Too often when natural resource problems arise, people jump immediately to recommend solutions. Of course, the very aim of management is to solve problems that threaten sustainable use of resources. But multiple solutions may be promoted, some of which may conflict, cause new problems, or offend some people. Even if people agree on a solution, it may address only parts of the problem or not actually solve the problem. Being conventionally "solution minded" rather than effectively "problem minded" means that we tend to make assumptions about people's goals, pay too little attention to what has happened in the past and what might happen in the future, and focus uncritically on possible solutions. As a professional, you must be aware of these limitations and know how to avoid them.

These difficulties and others could be avoided if participants focused on analyzing and understanding problems fully before they proposed solutions. This all-important step can help you and the people with whom you work invent better solutions. This problem orientation forces you to be realistic in your recognition of the scope of the problem and its varying contours of complexity and tractability, and it allows you in the end to be practical and effective in its solution. The social and decision process mapping examined in the previous two chapters is an "indispensable preliminary" to problem solving, but it does not supply you with any answers—only "a guide to the explorations or questions that are necessary if specific issues are to be creatively dealt with" (Lasswell 1971a, 39). It is by using problem orientation, the third dimension of the policy sciences' framework, that specific problems are analyzed and workable solutions devised.

This chapter examines the five tasks of problem orientation, each of which must be carried out to address problems comprehensively. It also examines how problem orientation can aid decision making and how problems are defined. Finally, two case studies are presented in order to illustrate application of the tasks.

A Strategy for Orienting Oneself to Problems

Many professionals do not think rigorously and reflectively about the method of problem solving they employ (see Schön 1983). For example, one Forest Service supervisor recently said that, although she could not describe to herself, her staff, or other people how she addressed problems, she was confident she was making good decisions. She felt that if she thought about it too much it might interfere with her problem-solving abilities. Some professionals, on the other hand, know explicitly and coherently the method they use, are able to describe its elements and procedures, seek to improve their skill in applying it, and are open to even better approaches.

In order to address problems realistically, a strategy is required. Rational action is the best way to proceed, regardless of the problem or the field of your work or study. "A man's action is purposively rational if he considers the goals, the means, and the side effects, and weighs rationally means against goals, goals against side effects and also various possible goals against each other." This is Max Weber's definition of rationality (cited in Reisman 1981c, 172). McDougal et al. (1981, 288) agree that "rational and realistic behavior is an interaction between explicit or implicit goals and a constant flow of information about the environment, directly from without as well as indirectly from 'memory loops'" inside a person's head. A policy is rational if it realizes the desired goals; rationality is determined by the outcomes of the decision process, not by people's intentions. Achieving perfect rationality is impossible, but it is an aspiration worth striving for; the alternative, irrationality, will not, by definition, serve the common interest.

The principal challenge to policy-oriented natural resource professionals is to map the social and decision process in ways that are relevant to carrying out the five tasks of problem orientation (see fig. 1.1; Lasswell 1971a). The following five sections examine the problem-solving tasks, drawing on Lasswell and McDougal (1992, 725–1128; see table 5.1). These intellectual tasks, which can help you address any policy problem, act as guides to both content and procedure. That is,

Table 5.1. An Overview of the Five Intellectual Tasks of Problem Orientation and Some Questions to Ask in Carrying Them Out

Tasks	Questions to ask
Clarifying goals	What goals or ends, both biological and social, does the community want? Are people's values clear?
Describing trends	Looking back at the history of the situation, what are the key trends? Have events moved toward or away from the specified goals?
Analyzing conditions	What factors, relationships, and conditions created these trends, including the complex interplay of factors that affected prior decisions? What models, qualitative and quantitative, might be useful at this stage to explain trends?
Projecting developments	Based on trends and conditions, what is likely to happen in the future? Project several scenarios and evaluate which is most likely. Is this likely future the one that will achieve the goals?
Inventing, evaluating, and selecting alternatives	If trends are not moving toward the goal, then a problem exists and alternatives must be considered. What other policies, rules, norms, institutional structures, and procedures might move toward the goal? Evaluate each in terms of the goals. Select one or more.

Source: Laswell 1971a; Clark and Brewer 2000.

they remind you to ask particular questions about any problem and to ask them in an orderly manner. Explicit reflection on how problems are understood and how solutions come about can have great impact on improving the performance of government, business, and civic and conservation groups as well as individuals in their daily lives (Maier 1962; Moore 1968; Clark et al. 1996).

Clarifying Goals

The first task is to clarify the goals of the participants. People involved in a resource management issue must specify what they hope to achieve (a content matter) and also how they expect to achieve it (a procedural matter). The obvious difficulty, of course, is that members of a society are rarely of a single mind about ends or means. Every group and every individual has multiple and often conflicting goals and makes varied demands. It is the job of the professional to help identify the claims actually being made and those that are valid, to find common elements among them, and to assist people—amid their competing demands—in acknowledging their shared goals and building workable, integrated policies from that common point. An environmental group may find, for example, that the only thing it has in common with a municipality in the case of land use planning disagreements is a commitment to a public forum to work out their differences. Or the two groups may discover that they share, along with other groups, a desire to diversify the community's support base or find better ways to work together in the future. With concerted effort, they may find that they share a number of fundamental values or goals, and from there they can proceed through the other steps of problem orientation to come up with alternatives. Goal statements should be as inclusive as possible in dealing both with substance and systematic proceedings in the social process.

Goals are the preferred outcomes in a specific context in terms of the distribution of values, practices, and institutions. Some practitioners may tend to think of goals as conventional, concrete management or conservation objectives. As I have stressed, however, developing a policy orientation calls for attention to the decision-making process, how people interact and participate in it, and its impacts on people's lives. For this purpose, looking at decision processes as transactions in values is a practical frame of reference that can be employed in all situations, no matter whether the issue is fisheries regulation, prairie restoration, or international development projects. Clarifying your own values and goals as well as those of other participants can best be done by being explicitly sensitive to these matters. Try to cultivate the habit of reflective, analytic insight. Some people avoid creative exploration of their own or their community's values through various anxiety-preventing defense mechanisms, often uncritically accepting conventional values embedded in society and developing rigidity in their thought and action. Psychologists tell us that whether life has meaning for people depends on whether they have goals that are centrally related to their self-concept and self-esteem and that express their current values.

Clarifying values means developing answers to the question "What value outcomes should we seek in a given set of circumstances?" It is useful to begin with choices that are fairly broad and widely accepted—again, stated in terms of social process—and then move to greater specificity. Lasswell and McDougal (1992, 737–58) recommend embracing as the overriding goal human dignity or democracy, which they detail in light of the eight value categories. This may strike some as a peculiar, even irrelevant, goal for resolving problems in the field of natural resources. But, again, we are looking at the effects of human decisions on human lives, which includes the health and use of the environment. Because of the sometimes extreme difficulty of finding shared goals or values in specific contexts, participants must step back from their individual demands, claims, and special interests to find common ground in more general, widely shared values. Sometimes this requires retreating to the most basic cultural myths and doctrines, the most basic human rights, or the most fundamental precepts of our nation or culture. This holds true for natural resource problems as well as any other arena where people disagree but must find common ground in order to live together. The central question is whether a given decision process serves to realize democracy and human dignity or caters to the select few. The Universal Declaration of Human Rights recognizes these values in calling for "life, liberty and security of persons," "right to marry and to found a family," "freedom and respect," "freedom of opinion and expression," "right to work," and "freedom of thought." The goal of democracy, well rooted in many cultural traditions throughout history, is for all people to have full opportunity to shape and share power, wealth, enlightenment, well-being, skill, affection, rectitude, and respect. One measure of democracy is whether all of society's institutions are accessible and open to appraisal to see how they function in achieving this overriding goal. We will come back to the notion of human dignity in relation to natural resources in the last chapter.

Having a general goal is vital, but to solve problems we need more detail. We must find ways to operationalize the overriding goal in given contexts. In a small, forest-dependent community, for example, where many residents' livelihoods are threatened by cutbacks in timber harvests because of endangered species regulations, competition from international firms, slowdowns in the building industry, or other reasons, we need to decide what courses of action by the community, the timber industry, the government, and other players would be in the community's best interest. Are residents most concerned about making a living, and would they be willing to accept help in developing new industries? Do they fundamentally lack respect, and if so, how could the rest of the participants show deference to them and their way of life? You should ask these kinds of questions about all eight values for all participants. In your analytical role as a professional, you should specify an appropriate level of detail in order to be practical in achieving the goals, you should specify concrete objectives in order to work toward the goals (part of the alternative-developing stage of problem orientation), and you should develop operational indices, focusing on short-, mid-, and long-term scales, in order to tell if the goals are being met.

In clarifying goals we must also be clear about procedure, the activities we will undertake and the order in which we will do them. Our society commonly overemphasizes content at the expense of procedure, but the five intellectual tasks can be a useful guide to the order in which substantive matters should be brought to our attention. Using this approach consciously and comprehensively, yet selectively, serves as a continual preparation for problem solving.

Finding an answer to the simple question "What do people want?" or "What should I do?" is, without a doubt, arduous. Some people avoid asking these questions at all and instead simply make assumptions about people's goals. Some people fall back on philosophy or religion to formulate their goals. In solving problems, however, policy-oriented professionals do not juggle competing claims and goals based on "transempirical" propositions. They use their skills to determine perspectives and events in social process empirically and do not allow their own beliefs, values, goals, or identifications to interfere with their scientific integrity and social responsibility.

Describing Trends

The second task in problem orientation is to describe past and current trends. This task may appear straightforward, but it can be complex and it is always important to do it well. Looking at things from a historical standpoint tells us how closely past events and decisions have approximated the community's goals, what discrepancies exist, and how great they are. Without this kind of knowledge, it is impossible to be rational because we cannot know where we are headed or what choices to make. In solving natural resource problems, we need to document not only trends in the status of environmental variables but also trends in the social process—whether the eight values are becoming more abundant and widespread in the community and whether institutional practices more closely reflect the goals of democracy and sustainability. Trends in people's perspectives are an important variable to map. Are they becoming more universal or more parochial? Is there wider support for the goal of human dignity, or is there movement toward tyranny and deprivation?

One global trend of major import, especially to those in resource management fields, is the changing relation of people to the environment. The human population has grown from a few million about ten thousand years ago to somewhere between six and seven billion currently. It may exceed ten billion by 2050. During this period the natural resource base has stayed relatively constant, and resources—plant and animal species, topsoil, clean air and water, the ozone layer, and scores of others—are being depleted or polluted. In addition, as we gain more knowledge, our relation to nature changes. Our demands have differentiated and so have our uses of the environment. As a result, the environment is constantly changing in its importance as a supplier of raw natural resources, both renewable and nonrenewable. During this ten-thousand-year trend, the attainment of human dignity has also changed. Different times and different countries have met the growing demands for values more or less effectively through their own systems of public order. Lasswell and McDougal (1992, 803–864) discuss this ebb and flow in terms of the growing demands for value sharing and the extent to which people have been able to realize their values. This is a complex accounting, especially in the twentieth century with the rise of the modern industrial states, various democratic and totalitarian nations, and world wars, lesser wars, and other breakdowns in public order. Today there are more demands for sharing wealth, knowledge and education, power, health, family and community integrity, respect, skill and work, and rectitude than ever before. Our natural resource base will be increasingly challenged by these demands. It is inconceivable that human rights will be achieved for all without simultaneously managing natural resources sustainably.

As you map trends, you should characterize the past factually, based as much as possible on reliable, verifiable (rather than anecdotal or impressionistic) knowledge. Describe trends in terms of social and decision process: How well have past events and decisions achieved people's goals? How well have they served the common interest? What were the consequences (outcomes and effects) of past events and decisions, and in terms of which values? Which groups, institutions, and practices were advanced and which pushed to the background? Mapping trends will give you a clear idea of whether a specific problem is getting better or worse, whether the trends are consistent with the community's stated goals, how far the participants have traveled toward achieving their goals, and where they have succeeded or failed. Determining trends is also affected by each of the other tasks in your problem-solving strategy: Clarifying goals tells you which trends to map, whereas analyzing conditions and projecting trends into the future may highlight the importance of one trend relative to another.

Analyzing Conditions

The third part of problem orientation is the analysis of the conditions or factors that affected past events and decisions. It tells us how and why the trends took place. This requires scientific examination. Once we understand trends and conditions, we can revisit our goals to see how realistic they are, and we can also make intelligent projections about what might happen in the future. Collectively, this knowledge gives us a means to evaluate policy alternatives or options.

The policy sciences' scientific approach to understanding conditions is distinctive. It involves formulating and testing hypotheses and developing models of the human enterprise and the resource environment to guide observers in what to see, how to see it, and why it is important. Without a comprehensive, practical, problem-solving theory we would be severely handicapped in understanding events and processes around us, and without a set of empirical methods to gather, process, and disseminate information we would be unable to investigate conditions.

Social scientists are in a unique position: they are "objective," outside observers of the social entity and processes of which they are also a part. The subjective status of scientists is important here, more so than when they study nature or some biophysical phenomena. They must rely on their own perceptions of data, some of which are subjective events that are not open to direct observation, such as emotions, feelings, thoughts, or patterns of identity. Many inferences must be made about individual, group, and institutional behavior. This requires scientists to undertake self-observation at the same time they observe others. Skilled reflection is essential (Schön 1983). Policy-oriented professionals undertake this observation in ways similar to those used to some degree by everyone in the course of everyday life. We all make generalizations about causes and effects in the interactions we see, and most of us distinguish between statements of preference (value goals) and statements of description, including cause and effect. In this sense, we are all scientists in that we observe and test the truth or falsity of statements and communicate our conclusions.

In the past four hundred years, science has become a dominant means of interpreting ourselves and the world around us. In earlier times, scholars such as Aristotle, Leonardo da Vinci, and Linnaeus used science in a descriptive sense, but today positivism is a predominant view that sees science as experimental manipulation of variables one by one to tease out cause-and-effect relationships. This conception is a restrictive, highly technical, and relatively recent invention that grew out of centuries of scientific outlook and the work of many individuals and institutions. Science as understood today is a coherent worldview or myth—with supporting doctrine, formula, and symbols—and an accompanying set of practices. Lasswell and McDougal (1992, 869) define it as a "systematic body of propositions verifiable by observation, all of which are not obvious to commonsense." In general, science "advances" by testing theories against empirical data derived from direct observation. It is a specialized skill focused on the value of enlightenment—its accumulation, application, and enjoyment. In human interactions science formulates, observes, and evaluates statements about social process that are descriptive, not preferential—that is, it considers what is, not what ought to be. It seeks explanation. In carrying out their work, scientists try to recognize and avoid their biases so that scientists anywhere can reach the same conclusions.

The purpose of the policy sciences, if not all science, is not prediction but improved freedom of choice (Brunner 1991a; Brunner and Ascher 1992; Lasswell and McDougal 1992). As science produces new knowledge about, say, the causes and conditions of a particular environmental policy, people can take that information into account and adjust their behavior accordingly. Similarly, new knowledge about the social process can change the focus of our attention, judgment, and actions and give us the opportunity to adapt ourselves or introduce new social routines. It increases our freedom of choice.

Making systematic observations about conditions is done using four scientific methods—cases, correlations, experiments, and prototypes—all of which seek reliable knowledge of conditions and trends. Case studies, which are usually qualitative and vary in detail, provide insight by giving an insider's view of events in a decision process. Correlation studies tend to be quantitative and statistical, which requires defining terms and determining probability statements about relations among variables. Experimental studies employ a design enabling observers to control all factors at will; factors can be varied individually or together as a means to find cause-and-effect relationships. Prototyping is an intervention intended to bring about a new arrangement of elements in a social process, the outcomes and effects of which are then studied. Prototyping does not share experimentation's high degree of control, and because it involves originality and uncertainty, confidence in prediction is not possible (see Chapter 7).

Some of the conditioning factors we might use to explain trends in ecology include geology, climate, and species interactions. In human social process the broadest categories of conditioning factors, used to account for the reasons events and social processes happen as they do, might include culture, class, interest, personality, and level of crisis or intercrisis. People behave in certain ways because of past and current experience with particular cultures and subcultures. Cultures may be global, national, regional, local or restricted to a city, a professional skill (for example, wildlife biology), or some other dimension. Class experiences also influence behavior. Although there is a tendency in the United States to deny the existence of classes, upper (elite), middle (mid-elite), and lower classes (rank-and-file) shape (accumulate) and share (enjoy) values in different ways. Upper classes have the most and lower classes the least of the values to be had in any social process. Interests also vary among people, cutting across cultures and classes. As a result, they may be more restrictive or more expansive than culture or class. Personality, the basic behavioral pattern people use to deal with self and others, is another way to account for individual behavior. Finally, the level of crisis (or intercrisis) is a condition affecting cultures, classes, interests, and personalities. Crises are events or processes that may have life-threatening outcomes and effects on people and institutions. People are not well prepared to deal with crises. Ultimately, the explanation for individual and collective behavior lies in social process—defined, functionally, as humans pursuing values through institutions using resources.

Inquiry into conditioning factors requires careful study. A notion of equilibrium underlies many people's thinking about trends and conditions in both human and natural systems. Analysis of conditions is the search for factors that move a system toward or away from equilibrium. Conceptions and theories to explain observed events and decision processes should be advanced and tested by appropriate methods of modern science. Here, as in the description of trends, you should look not only at problems in the natural world but also at events and decisions in the social process surrounding those problems. Your theories about conditioning factors should be based on the maximization postulate, which, again, says that people behave in ways that they perceive will leave them better off than will the alternatives. Behavior, as every biologist knows, is a product of many factors, and your inquiry into behavior should examine the interplay of multiple factors, including individual perceptions and other subjective events based on predisposition and environmental variables. Your efforts should be scientifically rigorous in analyzing conditions without overemphasizing theoretical models, mathematical measurement, or experimentation. Scientific quantification exists to aid judgment, not to supplant decisions.

Projecting Developments

The next task in problem solving is to make projections about future trends in events and decisions. The goal is to estimate the likelihood that important features of the social context, conditions, and problems will persist unchanged or how they may change in the future (Ascher 1978). You want to determine how likely it is that the community will realize its preferred goals. The first take on projection is extending past trends into the future with no intervention. If projections show that events and decision making are moving in the direction of achieving people's preferred goals, and at a satisfactory rate, then intervention may be unnecessary. But if projections show a future that is unacceptable (for example, a species' population will continue to decline until it becomes extinct), then something must be done, which is the fifth problem-solving task, inventing solutions. Trend mapping and scientific analysis look to the past, but since policy looks to the future we need to make projections about what is likely to happen.

Like the past, future events will depend on their contexts, so the principal challenge is to model our expectations of future developments as explicitly and dependably as possible. Like models used in ecological sciences, your projections of events and processes, including participants and their perspectives, changes in the situation, value dynamics, and possible decision outcomes, should be as clear, explicit, and realistic as possible. They should be subject to disciplined knowledge about conditioning factors and past changes. Simple linear or chronological extrapolations using commonsense, conventional theories are usually inadequate. It is too easy for people doing projection to fall victim to self-deception, wishful or fearful thinking, a battle psychology, and defensive policy. Systematic, disciplined, projective thinking is challenging. Two kinds of knowledge are involved: knowledge of trends and knowledge of conditions, both of which, ideally, are systematically organized. You should evaluate how future events will affect the shaping and sharing of values and institutional practices, continually moving back and forth between contextual features and the details derived from cases, correlations, experimentation, and prototyping. Most professionals and decision makers are very sensitive to future events and processes, yet they are uncomfortable with discussing them since there are obvious political hazards in being wrong. At the same time, being reliably knowledgeable about past trends and conditions, and therefore about the future, is a powerful aid in decision processes.

One important tool in projection is the *developmental construct,* which depends on various methods of inference to come up with a sequential pattern of likely future developments that can be tested against all available information. These constructs are often developed for whole policy arenas and large-scale social processes, although they can be used for smaller-scale processes, too. We could, for example, produce developmental constructs of the future of biodiversity protection in the northern forests or at national or global scales. Projection requires much judgment, and today forecasting and prediction are big business in government, in the private sector, and for many individuals.

Lasswell and McDougal (1992, 986–1031) provide a sweeping illustration of developmental thinking about the entire human enterprise. They present two contrasting constructs about the future, each with a very different system of public order in which the eight values are shaped and shared in dramatically different ways. At one extreme, trends and conditions may be moving us toward a "garrison-prison state" (dictatorship) in which a new order of rigid social stratification would be enforced by strict socialization, education, and policing activities. Throughout the twentieth century there were many developments and conditions indicating the likelihood of this situation. At the other extreme, events and social processes may by propelling us into a "universal public order of human dignity" (democracy), a possibility that is also suggested by many trends and conditions in the past hundred years. Depending on which construct dominates, natural resources will be used in very different ways. The sustainability of the resource environment will be a prerequisite for a lasting state of democracy and dignity. This calls for natural resources to be put at the service of the widest possible shaping and sharing of values, compared to the garrison state hypothesis, under which resources would be accumulated and enjoyed by coercive power elites. The paramount job for professionals, decision makers, and interest groups—as well as all of humankind—is to bias future developments strongly in the direction of a commonwealth of human dignity.

Inventing, Evaluating, and Selecting Alternatives

The final step in rational problem solving is to devise, evaluate, and choose policy alternatives. Although many people see finding a solution as the only activity that counts, policy-oriented professionals will see that this step cannot be carried out adequately without first clarifying goals, describing trends, analyzing conditions, and projecting trends.

Workable solutions will be those that bridge the gap between the goals that the community has identified and reality or probable future reality, or in other words, the current or anticipated state of affairs. Every effort should be made to encourage and foster creative exploration of problems and potential solutions. Work settings, teams, and institutions should be configured to bring out the very best creative thinking and judgment. For increased creativity, Lasswell and McDougal (1992) recommend alternating periods of concentration and withdrawal: after a prolonged period of concentrated work on a seemingly insoluble problem, participants should withdraw and turn to other pursuits. Flashes of insight may reveal a solution. Withdrawal to nature is often praised as a restorative and perspective-building exercise. Immersion in nature may put problems into perspective and stabilize our outlooks.

Three kinds of solutions to policy problems are possible: win-lose, compromise, and integration. In win-lose, "solutions" are found when the most powerful side wins at the expense of the losers. In compromises the contending parties are clear about the possibilities of value gains and losses and they work out a deal to minimize value deprivations. Integrative solutions are obtained when a new framework of cooperation is devised and adopted and new perspectives, practices, and frames of reference result. Integration involves genuine innovation in which the context is redefined, leading to situations in which parties generally get what they want and no one seems to have won or lost. Integrated solutions are not available for all policy problems, but where available they are clearly preferable to win-lose or compromise solutions.

After strategies to address the problem have been invented, they must be evaluated, and eventually a course of action must be adopted and acted upon. Each potential solution must be interrogated systematically in both general and concrete terms to evaluate its utility in solving the problem that has been identified and in achieving people's goals. Many techniques have been devised to do this. For example, "decision tree" analysis has been used in endangered species conservation and many other natural resource arenas (Maguire 1986).

One approach to evaluating alternatives has been proposed by Brunner (1995a). He recommends applying five perspectives to the arguments used to justify or ground alternatives—logical, substantive, comparative, procedural, and adaptive. The *logical perspective* focuses on whether something is missing from the argument. "A logically complete argument will make explicit the goals and expected consequences (trends, conditions, and projections) that justify a position for or against an alternative" (8). Present at least two alternatives—the status quo and something else (if only one or no alternative is given, there is little to evaluate)—and ask if others should also be considered. If the expected consequences of each alternative are not explicit, you should ask whether the alternative is realistic, specifically looking at trends, conditions, and future contingencies that ought to be taken into account. Make sure that the goals are clearly laid out in the argument for each alternative. If they are not, there is no basis to judge whether the alternative is worthwhile, and participants are justified in questioning the basis for choosing. Brunner warns that incompetent analysis or argumentation may cause logical fallacies. Sometimes the fallacy is intended, as in censorship, to hinder, obscure, or mislead. Sometimes arguments are designed to direct attention to only one alternative or goal to make it appear compelling. Some policies are constructed to mislead and control public opinion. Be on guard and always look for more rational policies and alternatives.

The *substantive perspective* focuses on the content of the argument. If the alternative is impracticable (for example, it cannot be selected for technical, political, or other reasons), or if expected consequences are unrealistic given trends, conditions, and projections, or if the expected consequences are unworthy in terms of the goals, then the argument should be rejected, along with the alternative it was designed to support. Consider the evidence and make substantive judgments based on experience.

The *comparative perspective* focuses your attention on other arguments that might be helpful in refining your understanding. Are there "significantly different alternatives, more realistic expectations, or more worthwhile goals in other arguments on the same issue, compared to the argument at hand?" (Brunner 1995a, 9). If so, then you can conclude that the argument is less than rational by comparative standards. Simply put, consider all sides of the issue.

The *procedural perspective* focuses your attention on how the alternative was obtained. For example, you should determine how information was acquired, processed, and disseminated. Find out if all the relevant information was considered in the process of developing the argument, whether relevant experts reviewed it, and whether they might have had conflicts of interest in processing it. Investigate whether censorship or some other promotional means distorted the information. If the procedure was faulty, then the alternative is less than rational.

The *adaptive perspective* focuses your attention on whether the alternative in question can be corrected if an error is found (in part or in whole) after action has already been taken. Can it be fixed? Brunner notes several barriers to adaptation. If the goals are obscure, for example, then it cannot be determined if the alternative is a success or a failure. If the necessary data are not available, people cannot tell either way. If accountability is too diffuse, no one can be held responsible. Correcting a chosen alternative may also be impossible because resources simply are not available.

Remember that an individual "makes these logical, substantive, and procedural judgments by examining and interpreting relevant experience" (Brunner 1995a, 10). In appraising proposals that have not yet been decided, we therefore need to consider the outcomes and effects, both intended and unintended, of earlier decisions of a similar type. Our collective experience can inform our judgments. For example, if a wildlife management agency has a consistent record of decisions about endangered species that are not rational in terms of the five perspectives just described, we should be cautious about its promises of better future performance.

Lasswell (1971a, 56–57) offers a set of questions that are useful in guiding invention, evaluation, and selection of alternatives. First, whose are the policy goals that are to be realized? Professionals or analysts need to be clear about their own identities, expectations, and demands as well as those of decision makers, interest groups, and all others involved in or affected by the decision process. Second, what is the problem or set of problems to be dealt with? Again, problems arise when people compare actual circumstances to their expectations. These discrepancies need to be probed in depth in order to define their scope and tractability. Third, what objectives is the policy process expected to realize? Objectives are set in reference to goals, but the underlying values of participants are often implicit, hidden, or inaccessible without systematic, insightful examination. Without explicit goals, participants may proceed in setting objectives based on ambiguities. Fourth, assuming that the specified objectives can be met, what is the chance that they will optimize the desired results? This requires careful analysis of possible decisions and their consequences in terms of near-, mid-, and long-range outcomes and effects. Fifth, what outcomes are most adequate to the desired effects? Who decides what and how? Remember that there are outcomes from each of the decision functions. Sixth, what predispositions are favorable, which are unfavorable, and which are neutral in reference to the outcomes as currently formulated? Will they be sufficient (assuming little effort to mobilize support)? People's predispositions or circumstances may enhance or hinder efforts to achieve outcomes as envisioned, and they may change in the future to be more or less favorable. Seventh, what strategies will result in the highest values? What is the probability of a successful outcome? Strategies should be considered in light of various contingencies and predispositions. The various instruments of policy (diplomatic, ideological, economic, and military) will be used independently or jointly.

In summary, your emphasis in problem orientation should be on identifying and understanding a problem fully in relation to the goals, formulating multiple alternatives, and assessing and comparing their likely effectiveness, their projected value outcomes and effects, their breadth in serving the common interest, and so on. This will require disciplined knowledge of trends, conditions, and future possibilities. Creativity is paramount. At some point, though, deliberation must stop, a decision must be made, backed by an assumption of responsibility by decision makers, and the chosen strategy must be implemented.

Problem Definition

Rational problem orientation is a major part of every professional's responsibility. One of your greatest contributions can be in helping to produce a realistic problem definition. Defining problems as discrepancies between goals and actual or anticipated states of affairs, as we have done, attests that problems are socially constructed, based on the perspectives and values of participants. As David Dery (1984a) notes, the very idea of problem definition indicates a constructionist, rather than an objective, exercise: problems do not exist "out there," they are not objective entities independent of humans, they are not found, identified, revealed, or discovered. They are defined. They are a direct reflection of the values and goals of different groups of people. Problem orientation gives us a problem definition. Without a clear definition, there is no basis for even talking about solutions, much less choosing and implementing them. Converting a complex, problematic situation into a clear, practical problem definition can be challenging, but this approach offers a successful method.

Defining a problem involves more than just finding someone or something to blame for a difficult situation. It is really about the social significance of a situation, its meaning, implications, and urgency. Problem definition "has to do with what we choose to identify as public issues and how we think and talk about these concerns" (Rochefort and Cobb 1994, vii). By focusing attention on certain issues or ignoring or neglecting others, we declare what is at stake. The defining process occurs in several ways, but it always has significance for an issue's political standing as well as its solution. Problem definition is a central topic of policy research and practical problem solving today.

David Dery offers a substantial and sophisticated treatment of the concept in *Problem Definition in Policy Analysis* (1984a). He indicates that often decision makers are not clear about what they want, so it becomes the job of policy analysts—in our case, policy-oriented natural resource professionals—to clarify matters. Often multiple problem definitions, some adequate and some not, compete for attention and resources. Dery recommends using a "qualified relativism" to judge their worth.

There are many obstacles to adequate problem definition. Chief among them are the parochial perspectives of professionals, analysts, and decision makers. Epistemology, personality, and other biases also come to bear. Organizations, which play an enormous role in our society and have a limited capacity to learn from experience, can be especially powerful in shaping problem definitions (March and Simon 1961). For example, there are cases in which decidedly limited problem definitions have nearly crippled endangered species recovery programs, increased the dominance and power of the government agencies that promoted these weak definitions, and contributed to further risk to the species (for example, Miller et al. 1996; Clark 1997a). As Daniel Katz and Robert Kahn (1966, 489) note, "the facts of organizational life often preclude the recognition of dilemmas and their requirements for a radical restructuring of the very basis of the problem." A dilemma is a problem that cannot be solved because it has been inadequately defined, but organizations do not and cannot see this because of their biases. Some inadequately defined problems nevertheless receive strong institutional backing because of problem "ownership," or vested interest in a certain definition. A participant (individual or organizational) who owns a definition and dominates a policy arena also dominates how the problem is conceived and acted upon. Such an agent tries to force on society a particular characterization of the essential causes, consequences, and solutions of the problem. Problem ownership is thus an issue of power and, especially for institutions, control.

Perhaps the best single account of the significance of problem definitions is that of Janet Weiss (1989), who defines problem definition as a "package of ideas that includes, at least implicitly, an account of the causes and consequences of undesirable circumstances and a theory about how to improve them" (97). She explains that problem definition serves three functions: first, as the "overture" (the start-up phase) to policy making, second, as the very process of policy making, and third, as a policy outcome. As overture, problem definition functions as an initial analytic framework for talking about a problem and studying it. As process, it constitutes the dynamics of mobilizing support, shifting consensus, and taking action. As outcome, it creates programs, actions, and language for talking about problems, people, and social processes in subsequent policy cases. These roles exert differing influences on the policy process. Definition is a central task in problem solving, and it is vital for professionals and decision makers to distinguish among the problem definitions that are in play in any policy process, how they came to be defined, and how they function.

Carrying out Problem Orientation

Mapping social and decision processes and knowing about the recommended standards for their adequacy are essential to problem solving. Yet this information only provides the context of the problem, which itself must be defined by attending to the five tasks of problem orientation. Whether professionals are aware of it or not, they always employ some form of problem orientation in their attempts to comprehend the scope and significance of problems in preparation for devising solutions. Some approaches are sketchy, glossing over analysis of the problem in favor of finding solutions and often failing to evaluate alternative solutions, instead jumping to promotion of one based on special interests. The challenge for you is not to skip or neglect any of the five tasks, which are a logically useful way of devising optimal solutions to the problems that stand in the way of people achieving their goals. This can be hard, especially when you are working with conventional problem solvers who tend to overlook some of the tasks. At its best, problem solving involves the five tasks interactively along with the context of social and decision process. This information must ultimately be integrated into an overall understanding of the goal, the problem, its context, and the alternatives.

A Method for Orienting Oneself to Problems

One way for you to hone your skills is to read papers in professional journals with problem orientation in mind. Many journals publish papers about natural resource management, usually describing a problem, analyzing causes, and recommending solutions. One issue of *Conservation Biology*, for instance, contained articles on bird communities in relation to farming practices, the use of pesticides to conserve rare plants, developing conservation awareness through pond restoration, and pollination disruptions worldwide, as well as a special section of eight papers on creating better policy and management decisions by explicitly analyzing uncertainty. Many reports, recovery plans, and other planning documents also discuss problems and propose solutions. Learn to read these analytically and functionally, asking which of the five problem orientation tasks are addressed and emphasized and which are omitted or superficially treated (see table 5.1). This exercise will give you the ability to understand the structure of the paper, the analytic approach taken by the authors, and the quality of their problem orientation. Some of the questions you should ask about each paper are the following:

Is the goal explicit or implicit and assumed? Is it clearly stated? How do the authors identify the goal? What you think the goal should be, if different from the authors' statement of it, based on your knowledge of the policy sciences? (Goals may target process and content improvements.) What is the right goal for the context as described? Why?

What trends are listed? In addition to biological or ecological factors, do the authors describe what has happened in the social and decision processes in development of the problem in relation to the goals? How many trends are identified? What data are given to characterize each trend? Have some obvious trends relevant to the goal been left out? What might they be? Are some trends moving events away from the goal and other trends moving toward the goal? Remember, the trends do not constitute the problem.

Do the authors analyze the conditions under which each trend is taking place? Do they note the conditions or circumstances that influenced events as well as determining cause-and-effect relationships (to the extent that it is possible to determine precise relationships)? You could set up a simple table of three columns labeled "trends," "conditions," and "projections." For each trend identified in the paper, and other trends that may be relevant but were left out, there should be a description of conditions accounting for it and a projection of future developments. If trends and conditions are well documented, then what is likely to happen in the future for each trend? Do the authors address trends, conditions, and projections adequately?

If the "futuring" task shows that projected events are likely to move social and decision processes further away from the goal, then a problem exists. When goals are clarified, trends known, conditions determined, and projections made, there is a strong basis for a realistic problem definition. How is the problem defined in the paper you are reading? Is it the right problem? If not, can you offer a more realistic one? Do the authors define the problem in relation to the goal?

Finally, what range of solutions or alternatives is recommended by the authors? Is it clear that the proposed solutions will actually solve the identified problems and achieve the goal? Have the authors evaluated and compared more than one alternative? Too often papers advance only the authors' single preferred alternative, which is typically a recommendation to conduct more positivistic research. Are the alternatives evaluated in terms of rationality, practicality, and morality? How was the evaluation done? Was it thorough? In your judgment, was enough creativity brought to bear in the problem orientation, especially in the invention and evaluation of alternatives? What could or should have been done to strengthen the authors' problem orientation? One of the most difficult questions is when to stop deliberating, make a choice, and act. Being committed to the goals and fully problem oriented and contextual puts you in the best position to judge when to select an alternative. Is it clear in the paper you are examining that the participants have adequately attended to the prior tasks and have arrived at the point of selecting an alternative? Do they do so in a form that will be useful for decision makers? Do they seem to believe that their job is over once they have made their recommendations, or does it seem likely that they will continue to accept responsibility for their selection?

As you become more skilled, these and related questions will easily come to mind as you read or review professional papers, reports, plans, proposals, and other documents. With practice you will be able to pick out the basic structure and arguments in written or oral presentations quickly, incisively, and fully. You will learn to ask questions, offer suggestions, and contribute to ongoing policy processes based on your problem-oriented, contextual assessment in order to make genuine improvements in finding and solving significant natural resource problems. Orienting yourself to problems in this way, sensitively and contextually, will greatly enhance your professional life.

Koala Management Planning

Our first example of problem orientation is a 1995 analysis of the state's koala management efforts in Victoria, Australia, in which I participated. The Department of Conservation and Natural Resources (CNR) had previously contracted for a koala management plan, which was produced in 1989. But because it was still a draft, it carried no authoritative signature, there was no formal program to carry out its recommendations consistently across the state, and there was growing criticism of the state's conservation practices with regard to the species. The goal of CNR was to be in a position to manage the species in accordance with its mandates. There were obvious benefits, then, to moving forward with a final plan that was technically sound and well supported by the public. A formal plan could enhance and coordinate koala conservation and rationalize management across the state, assist in development of national policy, probably position the state to take a leadership role nationally and internationally on this issue, and help divert some criticism. Writing a final plan would require further discussion with key people, interest groups, and the public within and possibly outside the state—"friends" groups, researchers, and other special interests. Diversity of input would help clarify goals, ensure that all important issues were addressed, and build broad support for the final management plan and its successful implementation.

In summary, CNR was conducting a problem orientation exercise. In previous years biologists and managers had formed some sense of the trends in koala conservation, conditions that had enabled or caused these trends, and the likelihood that major problems would develop (although there were different, but not divisive, perspectives among CNR staff about definitions of the koala conservation problem and what to do about it).

The 1989 draft plan had assumed goals without actually stating them. Without an explicit goal, it was unclear to the public and managers just where the state's management was to be directed. It was also impossible to evaluate CNR's performance because there was no basis for knowing which trends, conditions, and projections to examine or for deciding whether to continue with existing management practices. The draft plan had listed a number of hardships for koalas, including road deaths, injuries from domestic dogs, diseases and parasites, and habitat loss. But it gave no contextual assessment of whether the situation was moving closer to the assumed goals or farther away and no in-depth analysis of causes and conditions. It failed to address matters of social process, which are particularly important for a species that attracts such strong sentiment worldwide. The draft plan listed possible management actions, such as stopping road deaths, reducing losses from dogs, and ending habitat loss, without evaluation of their political and practical feasibility, ramifications for other land uses (for example, logging, suburban development, and transportation), or their adequacy in achieving the goals. Without clear goals and knowledge of trends, conditions, and projections, it was impossible to decide what the problems really were, which were most important, and what to do about them, when, and how.

I was asked to aid the department's efforts to explore management approaches that would best meet the state's needs and address the problems. My contribution was to organize a team effort (the draft plan had been produced by a single individual) that would explicitly make the final plan more problem-oriented and contextual in content and procedure. I worked closely for several weeks with the departmental biologist and other professionals who were knowledgeable about koalas (see the appendix). We clarified goals, offered a four-part problem definition, laid out four alternative solutions, and then justified our recommendations. Because the koala is a high-profile species throughout Australia and the world and is part of our common natural heritage, it was required that it be managed under the highest standards of sustainability and in an exemplary way. Well-organized social and decision processes would be essential to produce a management plan that was rational, politically practical, and acceptable to the broad community. In other words, our analysis mapped the koala problem and suggested what to do about it.

The group outlined a number of goals based on sound conservation biology principles: (1) koalas should be well distributed throughout Victoria in all existing suitable habitats or ecosystems and throughout their historic range to the extent that this is known and suitable, (2) some populations should be large and dense and others should not, and (3) populations should be connected genetically in varying degrees because of varying degrees of habitat connectedness. Although these and other goals identified in our analysis were met to some extent by existing koala distribution, abundance, and management, we felt that goal statements of this kind should be included regardless of current conditions. It could only be beneficial to koalas and helpful to responsible management authorities to specify goals and clarify standards or operational indices by which to determine if goals have been met. Additional goals included the following: koalas should be managed as "independent," viable populations in areas where populations are large and dense (about ten such areas currently exist); population sustainability should be ensured; koalas should be available for public viewing; koalas should provide a focus for public information, involvement, and review; and the species should be managed as a cultural resource in a cost-effective way consistent with public values. These goals might be refined or amended in the development of the actual management plan.

We identified a number of social, organizational, technical, biological, and policy process problems that stood in the way of achieving these goals. This was our problem definition, and we recommended that these issues be discussed and evaluated further in order to refine and update it. The chief social issue, for instance, was that the public was genuinely interested in koala conservation and deserved assurances that koalas were being managed sustainably according to best professional standards. We identified certain organizational, technical, and policy practices that were expected to meet this concern. Biologically, even though the koala did not currently appear to be threatened across the state, some local populations might have been in danger of extirpation because of habitat fragmentation, subviable status, and other reasons. We called for existing data to be better organized and more accessible and for gaps in data to be filled in order to facilitate more precise assessments of koala status and management effectiveness at state, regional, and local levels. The data irregularities evident at that time were remedial and were expected to be addressed. We conducted an analysis of trends, conditions, and projections with regard to these problems.

Finally, we detailed and evaluated four possible responses to the koala conservation problem as we defined it: (1) to stop translocations and cease all active management of koalas; (2) to maintain the status quo, continue with existing commitments and efforts, and write no new final management plan; (3) to devise a new management plan that reflected existing management practices and views and took a science-based approach to technical and biological issues; and (4) to come up with a new management plan that reflected a practice-based approach and simultaneously addressed social, organizational, technical, and policy dimensions. For various rational, political, and moral reasons, which we detailed in our analysis, we recommended alternative (4). The recommended alternative is a departure from past approaches to koala management in that it would explicitly clarify current management standards as a basis for improving them and would address previously neglected social variables. We detailed an outline for a new management plan.

This analysis figured into the state's evolving policy stance on koala conservation. Shortly after our review, an interstate effort to coordinate conservation throughout the country was organized by the federal government, and Victoria's problem-oriented effort was fed into the national policy debate (Cork et al. 2000). Australia is still seeking an effective koala conservation policy while at the same time implementing some measures that can help conserve the species.

Banff–Bow Valley Task Force

The world-renowned Banff National Park in the Canadian Rockies, a region of vast forests and towering, snow-capped mountains, is Canada's first national park (Banff–Bow Valley Task Force 1996). It surrounds the Bow Valley with its clear, cold, glacier-fed river. The future of this park and region has been in question because of certain developmental trends and conditions that are degrading the environment. The government set up a task force to address current and future problems in this area (although these were not described exhaustively or contextually in the report). The broad goal of the Banff–Bow Valley planning exercise was to ensure the integrity of this region in perpetuity. The park's goal is to manage the park for the benefit, education, and enjoyment of the people and to keep the park unimpaired for future generations, as noted in the 1930 National Parks Act.

The task force was charged with three assignments: (1) to develop a vision and goals for the Banff–Bow Valley region that would integrate ecological, social, and economic values; (2) to complete a comprehensive analysis of existing information and provide direction for future collection and analysis of data to achieve goals; and (3) to provide direction in the management of human use and development in a manner that would maintain ecological values and provide for sustainable tourism. The team of highly experienced professionals took this charge and focused on four trends: legislation and policy, attitudes and public opinion, human use, and ecological integrity. For example, the analysis of trends in human uses showed that the park was greatly affected: in the mid-1990s, about five million visitors came each year, two towns were situated in the valley, and a transcontinental railway, a four-lane highway, and three major ski hills were sited in the park and the valley. These features and their ongoing impacts plus expected growth threatened the park's and the valley's ecological integrity. The task force concluded that if allowed to continue these trends would cause serious, irreversible harm to Banff National Park's integrity and its future value. Such diminishment, they noted, would weaken its attraction as a tourist destination and the associated contribution to the local, regional, and national economies. Since Banff tourism contributed six billion dollars annually to the economy, alternatives clearly had to be found.

The task force went about its work vigorously. Many meetings were characterized by frank, often spirited exchanges. Much information was marshaled and diverse perspectives put forth. Demands and claims were made and justified in various ways by opposing participants. Many mechanisms, such as the Banff–Bow Valley Round Table, were employed to produce a large technical report as well as a summary document. The task force acknowledged that its understanding of the complexity of the issue was incomplete, yet it had to come up with alternatives and so offered principles for Banff's management. After two years of deliberation, their report, "At the Crossroads," acknowledged that Banff faced some crucial decisions. The general strategy to solve the identified problems was based on what team members called the "principle of precaution." Deciding that a clean and abundant water supply and a healthy, complete, naturally functioning ecosystem (with its dependent tourism industry) were too valuable to lose, they recommended that developments and decisions that jeopardize Banff and Bow Valley be rejected or put off until more is known. They called for a new path, new goals, and the means of achieving them. Scores of detailed recommendations focused on ecological integrity, tourism, human use, management, basic and essential facilities, communities, commercial enterprises, transportation, and regional and park management—in other words, changes in ordinary decision processes and substantive or technical matters.

The recommendations, however, also included a "key components of governance framework," which was presented in order to bring clarity and better answer the "governance question" (63). The report noted that there was a great deal of concern about governance and decision making by government and by society at large. The framework outlined the task force's thinking about the nature of governance and ideas about reform. The list of framework components included "the current governance model, organizational initiatives, . . . planning processes, decision-making and accountability, public involvement and communication, data management" and other factors (63). In essence, the task force sought to improve the constitutive policy process. Overall, the community, via the task force and its inclusive operations, dealt with an extremely complex task of sorting out conflicting goals, documenting important past and future trends, identifying the conditions of these trends, and recommending a broad range of solutions to decision makers.

Conclusions

As they interact in social process, people perceive problems as discrepancies between their existing or anticipated situations and the way they would like things to be, that is, their preferred goals and value outcomes. The process through which communities achieve consensus and find enduring solutions to their problems is the decision process. It is vital, therefore, for those involved in problem solving, which includes professionals, authorities charged with making decisions, special interests, and ordinary citizens, to have a framework that permits them to explore problems fully in their social and decisional contexts. If a problem is not understood well, no amount of effort in changing the social and decision processes can resolve it. Problem orientation brings many issues into sharp relief as problems are considered systematically and in detail. The five intellectual tasks (clarifying goals, describing trends, analyzing conditions, projecting trends into the future, and inventing, evaluating, and selecting alternatives) are guides to both content and procedural considerations. Carried out sequentially, interactively, and iteratively, the five tasks constitute a very practical skill for helping people to assess whether their proposed solutions can and will actually solve their problems.

Problem solving is key to achieving successful decision outcomes. The way to clarify and secure common interests is to identify problems realistically and solve them in ways that are reasonable, practical, and justified. With comprehensive, contextual problem orientation and good judgment, many policy problems can be defined and solved.

SUGGESTIONS FOR FURTHER READING

Dery, D. 1984. *Problem definition in policy analysis.* Lawrence: University of Kansas Press.

Lasswell, H. D. 1971. Problem orientation: The intellectual tasks. Pp. 34–37 in *Preview of policy sciences.* New York: American Elsevier.

Weiss, J. A. 1989. The powers of problem definition: The case of government paperwork. *Policy Sciences* 22:92–121.

6 Policy-Oriented Professionalism
A Unique Standpoint

All participants in decision processes related to natural resource policy and management have their own perspectives, base values, and strategies that they use to affect outcomes. This is also true of policy-oriented professionals, but these individuals seek to maintain a different kind of outlook, a special standpoint relative to the decision processes of which they are a part. Having a policy orientation means being interested explicitly in gaining knowledge and insight into the decision process in which you are both a participant and an "anthropological" observer. It entails using this perspective as well as knowledge and skills to aid other participants in finding common ground. In contrast, most policy participants (for example, advocates, business leaders, local officials, environmentalists, and conventional professionals) are largely interested in their own special interests, even though they may try to couch them in language about the common interest in an effort to justify their demands. In order to fulfill this special role and responsibility, policy-oriented professionals must be able to examine and understand themselves and others in the context of any decision process.

This chapter describes and recommends a "policy orientation" to professional practice, compares it with conventional standpoints, and looks at what is involved in establishing and maintaining a policy-oriented standpoint in terms of science, pragmatism, and functional outlooks. I then describe career paths open to policy-oriented professionals as they develop their skills, competently explore the policy process, and try to improve its content and process.

A Unique Orientation to Serve the Public Good

Policy-oriented professionals have a unique ability and responsibility to clarify policy situations for themselves, for their more conventional colleagues, for ordinary participants, and especially for decision makers (see fig. 1.1). "If the policy scientist," or, I might add, any policy-oriented professional, "is to play his clarifying role in collective or private circumstances, . . . he must fully understand his own position. To some extent we are all blind and no doubt will remain so. But there are degrees of impairment, and so far as decision outcomes are concerned, it is the responsibility of the policy scientist to assist in the reduction of impairment" (Lasswell 1971a, 40). As a natural resource professional, you are given daily opportunities to help decision makers and other people reduce blindness and impairment through, for example, carefully and thoroughly analyzing problems, mapping contexts, defining problems, and evaluating potential solutions—efforts that help to upgrade rationality, make politics more practical, and provide sound justifications. Whether you are able to capitalize on such opportunities depends in large part on your standpoint.

Characteristics of a Policy Orientation

Like all policy participants, policy-oriented professionals must carry out rational inquiries and make rational decisions, and to be effective they must also have some influence on decision functions, especially intelligence, promotion, and appraisal (Lasswell 1971b). But they need to distinguish themselves and their purposes from the decision process under observation and from the purposes and methods of ordinary participants by being able to carry out dependable, realistic, and effective inquiry. In order to establish and maintain this knowledge-oriented standpoint, you must cultivate an understanding of the processes and events that allow you to carry out the tasks of problem orientation relative to social process (Clark 1992). Yet you must avoid being so theoretical or arcane that you cannot communicate, explain, and justify your perspective, role, analysis, and advice to other participants. If conventional understanding dominates your thinking, you will lose your unique standpoint and methods of inquiry, as well as the enlightenment toward which you strive.

All people have biases, but those with a policy orientation strive to recognize and minimize biases in culture, class, group, organization, discipline, epistemology, personality, previous experience, and other factors. In this way they differ in standpoint, role, intent, and action from other professionals or community members who make self-interested claims and from decision makers who must respond to those claims. This requires that they work to identify the various participants' common interests, which ordinary participants and decision makers may not be able to perceive for themselves.

Policy-oriented professionals ask questions of themselves that ordinary participants never address (Clark and Wallace 1999). The following are suggested by Andrew Willard (personal communication): What roles are you and other people engaged in while working in natural resource decision processes—scientist, technician, manager, student, teacher, advocate, advisor, reporter, decision maker, scholar, facilitator, concerned citizen, or others? What problem-solving tasks do you carry out when performing your roles—clarifying goals, determining historical trends, analyzing conditions, projecting trends, and inventing and evaluating alternatives? What factors shape how you carry out your tasks and roles—culture, class, interest, personality, discipline, organization, and previous experience? What conditioning factors shape your personal and professional "approach" in general and in reference to any particular policy and management case? Which approaches or roles are you predisposed toward or against, and how are you predisposed to conduct your professional work with respect to each? How does your approach shape how you carry out the tasks associated with your roles? For example, what is the impact of your "reflective, self-questioning approach" on the goals you clarify and how you specify them? the trends you identify and describe? the conditions you analyze and how you analyze them? the projections you make and how you make them? the alternatives you invent, evaluate, and select?

Probing these questions makes policy-oriented professionals unique; they simultaneously become part of, yet distinct from, the decision process. In practical terms, you should be an active member of the different communities with which you identify (for example, the wildlife society, the civic club, or the state game and fish department), but you should focus on the functional value processes at play in policy processes, on enlightenment (that is, increasing knowledge about the process and making sure it is widely available), and on your role in clarifying and securing the common interest within the community.

Toward a New Career Outlook

Professionals can come to a policy orientation by explicit examination of their own thoughts and behavior relative to the policy processes of which they are part or through years of experience. The first way requires a conscious effort, whereas the second way may or may not be fully conscious.

Professional careers take many forms. Some professionals, after years on the job, loosen the bonds of positivism and convention and acquire a pragmatic, functional approach on their own. Such people come to serve the public good through their analytic and social process skills, as opposed to merely promoting their own (or their employer's) special interests. Little by little, perhaps unconsciously at first, they develop their own vocabulary and partial, functional equivalencies to the concepts and terms we have learned here to describe what they do differently and more successfully than their co-workers. They may eventually become quite adept in applying these new perspectives and skills, even though they may be unable to characterize exactly what it is that they do or fully communicate their new operations to others.

Lasswell (1971a) identifies a clear path that individuals tend to take toward a policy orientation, noting that those who are free to define their own roles tend to come to it more easily. Practitioners generally start within a discipline and then magnify that discipline. Then they advance to related or other disciplines, which is what we see happening in forestry now, for example, in its adoption of remote sensing, economics, risk assessment, social valuation, and decision theory. This trend is a reversal of the specialization that has been common in education and practice (Barrett 1978). Next, as the trend continues these individuals reach out even further to contact diverse other people and knowledge areas. They move from an interest in explanation as a scientific task to an interest in achieving fundamental goals such as effective democracy and efficiency within bureaucracy. Finally, they seek to clarify and improve the social consequences and political implications of knowledge for the common good. This policy-oriented career represents a new group of skills and an expansion of the professional role to a much wider context. The result is a career that incorporates a distinctive integration and synthesis of technique. That is, it becomes interdisciplinary.

Many professionals become knowledgeable innovators or use sophisticated technical methods, such as computer simulation, surveys, or cost-benefit and risk assessment. But if they merely adapt their special tools to the social environment and are not truly interested in knowledge integration and synthesis or in clarifying and securing community interests, then they never make the leap to a full-fledged policy orientation. The challenge for modern professionals is to appreciate the career trajectory illustrated here, assess their own position, and accelerate their progress toward a policy orientation.

Serving Decision Making

Today, policy-oriented professionals work at levels ranging from the local to the international in ever-widening circles of "clients," that is, those who use their skills and knowledge, including bosses, co-workers, colleagues in other organizations, decision makers, interest groups, and the general public. As people better understand what such professionals have to offer and as the expectations of these users become more realistic, more and more will seek their help. Their knowledge and skills can be applied to the entire decision process, but they can be especially helpful in intelligence (planning) and appraisal (review) functions (Clark 1997b). The people in most private organizations and government agencies do not look at decision processes as a whole and as a result often misunderstand why they fail to unfold as hoped. This failure of understanding is usually part of the problem that must be resolved. In many instances, poor decision processes are also the product of badly designed and coordinated policy structures, although they are often misattributed to "the opposition," unforeseen events beyond the control of participants, poor public relations, or lack of resources. Policy-oriented professionals can help people understand these failures and avoid them in the future.

Professionals can aid clients by providing three indispensable appraisal services (Lasswell 1971a). First, they can help decision makers understand how well or poorly any particular policy process is performing and direct attention to the factors responsible. Second, and closely allied with policy appraisal, they can aid in impact appraisal by identifying the effects of organizations on social and decision processes. This work focuses on events internal to the organization carrying out a decision process. Third, they can help with constitutive appraisal, which examines policy processes to see if there have been major changes in the power positions of the people, groups, perspectives, or operations involved. Such changes in power, which come about from reform or revolutionary movements in social and policy process, would signal fundamental shifts in how all ordinary decision making would be carried out from that point on. The central question to ask in this kind of appraisal is whether changes in power relations have remained within some limits typical of that policy process or whether they have moved to a new state or a different arrangement.

This kind of analysis focuses on changes in authority (formal power) and control (effective power).

The policy-oriented practitioner can also be of great help in the intelligence activities of ordinary decision processes (Clark and Reading 1994). Without good intelligence, ordinary decision process devolves into a "muddling-through" exercise (Lindblom 1959). The job here is to mobilize knowledge from within and outside the framework of the established systems or institutions of authority and control. In your role as a professional you can help clarify goals, define problems, suggest creative solutions, and evaluate potential solutions. You can also find information that would otherwise be overlooked and people who could assist with the problem at hand and its context. A policy-oriented professional can facilitate other decision functions, too—enabling public debate and public relations, serving legislators or high-level decision makers, determining how to optimize use of scarce resources, specifying how best to apply policy prescriptions contextually, or helping authorities terminate a prescription and move on to its successor.

In order to be effective, professionals must be trusted by the people they are trying to help. Many people and organizations are aware that their participation in policy-making activities could be improved but are often unwilling to seek help, embrace it when it is offered, or trust advisors who are trying to be genuinely helpful. Your professional advice will be most effective when decision makers are open and willing to be fully examined. You will be most successful if you ally yourself with clients, co-workers, or decision makers who have complementary or at least compatible interests. Even then, helping people perceive their interests is difficult since people are not always fully aware of what their true interests are. A combination of common sense and formal policy analysis is required to bring these interests into the open. This may take you into a delicate exploration of the client's identities, expectations, and demands.

Policy-oriented professionals must be especially sensitive to the needs of decision makers. Demands from diverse interests, time constraints, complexity, and other factors can lead to enormous personal pressure on them, such that some resort to rhetoric about solving problems instead of actually solving them. Many communication problems can arise because of this. Their decision-making ability depends in large part on their receiving information in a readily usable form. Because they are often very busy, you may need to package your analysis and recommendations in easily consumable, simple messages. Considerable thought and effort should be put into the presentation of policy problems and options to decision makers. Simple charts and figures that convey complex analysis in easily understood terms are particularly helpful (Tufte 1990, 1992, 1997). Regardless of how decision makers deal with pressure, they generally realize that all policy problems have impacts that require expert attention and that specialized help is needed to bring reliable knowledge and judgment to bear (Brewer 1995). As a policy-oriented professional, you must be ready to meet this need.

Determining Your Standpoint

I have differentiated policy-oriented professionals from their more conventional colleagues because they explicitly establish and maintain an observational standpoint for themselves that serves the common interest—services for which there seems to be greater appreciation and demand in recent years. They have the ability to formulate problems, focus inquiry, explicitly postulate public-order value goals (human dignity), and carry out orderly problem-solving tasks. Table 6.1 compares how policy-oriented professionals and their more conventional colleagues understand the world and carry out their work. The two approaches to practice differ in three important factors: knowledge system (positivistic or post-positivistic), outlook and procedures (scientific or pragmatic), and understanding (functional or conventional).

Knowledge Systems: Positivism and Post-Positivism

Epistemology is the study of the mind's way of knowing. The ancient Greeks were keenly interested in epistemology (see Williams 1979). Perhaps their interest started when they noted that people in other lands differed in what they thought they knew. There seemed to be many apparently workable "world views" in existence across diverse cultures. How could this be, given that there is only one world? This led to epistemological questions such as the following: What is absolutely true? What can we know for sure? What is really here? What can be known? How can we know? How can we know what we know? Epistemological inquiry continues today in philosophy, cognitive psychology, and other fields. Researchers are examining the structure and physiology of human thought and perception in order to understand how the mind imposes order on experience and makes meaning from this order and to establish the relations between perceiving and thinking, between inventing and knowing, and between thought and action (for example, Bruner 1990). Interestingly, the more the subject of epis-

Table 6.1. Comparison of Two Professional Standpoints in Natural Resource Policy and Management

Conventional professionalism	Policy-oriented professionalism
Professionals follow a specified plan or project design (usually positivistic)	Professionals do not know in detail where projects will lead so work is an open learning process
Strong natural science bias	Mixes natural and social sciences
Scientific method is positivistic and reductionistic (goal: cause and effect, prediction)	Scientific method is holistic and post-positivistic (goal: human freedom, improved judgment)
Assumption of single, tangible reality (objective nature, positivistic epistemology)	Assumption of multiple realities that are partially socially constructed (pre- and post-positivistic epistemology)
Problem solving is blueprint-like, but empirical, systematic	Problem solving is process-like, yet empirical, systematic
Professional categories and perceptions are central to problem solving	Local categories and contextuality are central to problem solving
Understanding extracted from controlled situations	Understanding comes from interactions with context
Use disciplinary-based, acontextual, and limited methods	Use problem-oriented, contextual, and integrative methods
In practice, professionals control problem solving and clients	In practice, professionals enable and empower people in close dialogue about problem solving in context
Professionals often work alone with single disciplinary focus	Professionals often work in groups with an interdisciplinary focus
Careers are inward and upward	Careers include outward and downward
High-level professionals lose touch with changing local realities	Professionals stay in touch with action at all levels and dimensions of context

Source: Schön 1983; Pimbert and Pretty 1995; Clark and Wallace 1999.

temology is researched, the more difficult it is to put the questions to rest (Goldman 1986).

Acknowledging the basic nature of human perception and knowing will give you a more confident base of professional operation and permit you to proceed with problem solving in a more pragmatic way.

Different modes of problem solving are underlain by different episte-mologies and supporting ideologies, which, in turn, greatly affect how people think and act (see table 6.1; Bem 1970; Beardsley and Beardsley 1972). The epistemology of ordinary science is positivism, which is well established in most approaches to problem solving in natural resources and is entrenched in nearly all organizations. Positivistic science was designed to deal with certainties and, in modern times, probabilities. It is one common means by which diverse people can come to agree on what is real. It works well in the ordinary world as long as its users do not look at its epistemological foundations.

We now know from much epistemological inquiry, however, that positivistic science's measurements, even mathematics, are not abso-lutes that describe the "objective world." The works of Albert Einstein, Werner Heisenberg, and Kurt Gödel in physics show that it is impossi-ble to find a firm basis for knowing. The writer Annie Dillard (1982, 55) posed the key question: "What can we know for certain when our position in space is limited, our velocity may vary, our instruments contract as they accelerate, our observations of particles on the micro-level botch our own chance of precise data, and not only are our own senses severely limited, but many of the impulses they transmit are edited out before they ever reach the brain?" Even if we could have confidence in our senses, we may not be able to trust what our brains do with the input, and even if there is a basic structure to nature, sci-ence must work its way through linguistic and cultural assumptions to be understood, expressed, and communicated. Heisenberg con-cluded that we cannot study nature as it is, only our perceptions of nature. And Sir Arthur Eddington (1927, cited in Dillard 1982, 56) noted that "the physical world is entirely abstract and without 'actual-ity' apart from its linkage to consciousness." Most epistemologists have concluded that all knowledge is contextual and only contextual. Hu-mans cannot know an objective world (assuming one exists) derived as assumed from positivistic science. We can only share subjectivities, each person's individualistic understanding of self and the world. We can only compare notes and come to some kind of tentative, intersub-jective, contextual agreement about how the world seems to work (Ha-bermas 1987). This latter view of the world is the epistemology called *post-positivism.*

The practical challenge for problem solvers is how to proceed when nothing can be known for certain. The policy sciences' problem-oriented, contextual method was invented to improve your judgment in the face of the contemporary understanding of epistemology.

Outlook and Procedure: Positivistic Science and Pragmatism

I have just contrasted positivism with post-positivism in terms of how people know what they know. In terms of people's outlooks and procedures in solving problems, positivism also stands in contrast to pragmatism, a philosophy invented in the United States about one hundred years ago to address problems realistically. The policy sciences are rooted in pragmatism. With regard to our fundamental question of how to solve natural resource policy and management problems, we must ask whether positivistic science or pragmatism is the more practical approach.

Positivism, the philosophy behind modern experimental science, which underlies much of natural resource policy and management (for example, Miser 1956; Richards 1983; National Research Council 1986), is "the doctrine that causal laws of society," and, we might add, the biophysical world, "can be deduced from basic assumptions, verified by empirical test, and then added to a stockpile of inviolable truth" (Dryzek 1990, 33). Positivism assumes that facts simply are (Harre 1985) and are discovered through properly passive observation of natural reality. Positivistic philosophy is often accepted uncritically in part because, as William Sullivan (1995, 166) notes, it has been central to university education for more than a century and is reinforced through on-the-job socialization. This is certainly true in natural resource fields. "The continuing hold of positivistic dogma over the thinking and practice of higher education is a key problem which must be confronted by anyone who concludes that the needs of our time demand a reshaping of professional knowledge as well as the way professional life is organized" (Sullivan 1995, 166). Not surprisingly, most professionals draw on positivism automatically and unconsciously when they address problems and promote solutions. It is so broadly and powerfully accepted and used in human affairs (for example, conventional economics, Friedman 1953) that it has been taken as given and only occasionally has its utility in problem solving been disputed.

Positivism and its offspring, experimental science, are the targets of growing criticism, however. First, many people have questioned how well positivistic science has served society during the past century (for example, Blanco 1991; Jamieson 1991; Horgan 1996). It has provided many benefits—stunning advances in health care, transportation, and communication among them—that must be fully appreciated. It will continue to benefit humanity. But critics have demonstrated that fact-finding is not as objective as once thought but is really a human, value-laden social process (Fleck 1979; Berger and Luckmann 1987; Latour

1987). Moreover, positivism has also been shown to be the principal cause of numerous policy failures. Much of the current debate about its use in policy focuses on the objectivity and subjectivity of people (see Fish 1996; Sokal 1996). For example, John Dryzek (1990) concluded that many of the world's present political ills have much to do with positivism (or instrumental rationality, as it is sometimes called) and its use in policy, and he feels that these underlying thought patterns are one of the major problems of our time. Positivism destroys "the more congenial, spontaneous, egalitarian, and intrinsically meaningful aspects of human association. It is antidemocratic. It represses individuals" (4). The political and social institutions that manifest positivism are ineffective in resolving complex social issues, and their use precludes more appropriate policy instruments. He argues that positivism has also been inappropriately and ineffectively used in the social sciences in such methods as opinion surveys and that objectivism itself is a false and repressive problem that "inhibits the progress of political science and politics" (7). Dryzek says the cure involves large doses of what he calls discursive democracy, which seeks to combine democracy and problem-solving rationality, an idea originally suggested by Lasswell (1951a, 1951b) in his publications on policy sciences for democracy.

The second criticism of positivism is that it is an ideology in the physical and natural sciences (Kuhn 1962; Rose and Rose 1976; McCain and Segal 1977). It functions as a myth to justify a special place for science and scientists in society. It encourages narrow, problem-blind, acontextual, reductionistic, and single-method inquiries (that is, quantitative, formal-deductive, and predictive), and according to Dryzek (1990), positivism is the culprit behind bureaucratic hierarchy, much manipulative political science, and many coercive political practices.

Third, the goals and standards of positivistic science (see Friedman 1953), which include prediction with precision, scope, and accuracy, cannot be met in many real-world situations, and thus the use of positivism in various policy applications, such as experiments, complex predictive models, and rational choice theory (see Ascher 1981 and Simon 1985) has been critiqued. Because of this, Brunner and Ascher (1991) have noted that positivism is waning in importance although it is still quite powerful (see also Manwell and Baker 1979).

Pragmatism, on the other hand, is a philosophy designed to address real-world problems. It transcends ordinary perspectives, positivism, and technocracy, according to Torgerson (1985). The pragmatic approach to problem solving originated with C. S. Peirce and John Dewey (see James 1978; Dewey 1981–90), who criticized scientific positivism for ignoring the fact that all inquiry takes place as part of a social process and thus ignoring the goals implicit in any given act of social process. Pragmatism argues that knowledge results from inquiry, that inquiry is always a response to a "perplexity" that has disrupted a social process, and that all inquiry is directed by some end or purpose—a value goal. The purpose of inquiry depends on the character and identity of the investigator and the context. The pragmatic approach focuses our attention on the processes of knowing and inquiry, which are human activities and therefore part of social activity, and away from a narrow view of instrumental rationality and objectivism. Regarding the detachment presumed by positivism as false, the pragmatic approach adopts a more active, reflective stand toward knowledge and knowing (Sullivan 1995).

Donald Schön (1983) was one of the philosophical heirs to this tradition who researched positivism, professionalism, and problem solving in society. Looking at professional training in positivism, its consequences, and the relation between the kinds of knowledge honored in academia and the kinds of competence valued in professional practice, he became convinced that "universities are not devoted to the production and distribution of fundamental knowledge in general. They are institutions committed, for the most part, to a particular epistemology, a view of knowledge that fosters selective inattention to practical competence and professional artistry" (vii). Schön elucidated the influence of positivism on how professionals function in social and policy process, especially through bureaucracies: positivism's approach to expertise and problem solving, despite claims to the contrary, is "value laden, and technical experts have interests of their own which shape their understandings and judgments" (346). He concluded that new approaches must be promoted and called the single most promising approach "reflective practice," which is one element of what we call policy-oriented professionalism.

The positivistic and pragmatic approaches to professionalism can be compared in terms of their views of science, rationality, and politics in human affairs (Brunner 1996b; see also Brunner and Ascher 1992). First, they vary dramatically in their view of the purpose of science in society. Positivism holds that the purpose of science is to predict, with precision and accuracy, the consequences of a comprehensive range of possible changes in a system and that science can discover and use objective laws of nature for this purpose. Pragmatism, on the other hand, sees the purpose of science in human affairs as increasing freedom through insights that bring more factors more reliably into conscious awareness for purposes of decision making. Such freedom limits forecast accuracy. The two approaches also differ in their understanding of rationality in social affairs. The positivistic approach equates rationality with science. It holds that scientific predictions reduce uncertainty and are thus prerequisites to major policy decisions that are rational, comprehensive, and cost-effective with respect to fixed or given national goals. Pragmatism sees matters quite differently. Given uncertainty and ambiguity, it argues that it is more rational to field test multiple, discrete alternatives, to select, diffuse, and replicate the successes, and to allow for the adaptation of goals as well as alternatives. Finally, the two approaches vary in their view of politics in social process. In the positivistic approach, political consensus depends on objective scientific knowledge, and scientists deserve a privileged position above politics because their inputs to policy decisions are objective and value-free. In contrast, as Brunner notes, the pragmatist accepts that political consensus depends on trust and credibility derived from experience and that the values and beliefs of scientists, like those of other participants in politics, should be transparent to the public and their representatives.

Understanding: Functional and Conventional

The third factor in distinguishing policy-oriented professionals from their more conventional counterparts is that they understand social and policy systems in functional terms. As discussed in Chapter 1, someone using a functional standpoint looks for connections, relationships, and systems properties in social and decision processes, connections that are frequently overlooked by those who uncritically and unreflectively use ordinary ways of understanding, talking, and doing. Functional understanding explicitly depends on a comprehensive model, map, or image of social process to guide attention to the value significance of details (Lasswell and McDougal 1992). The policy-oriented professional sees the same events and processes as others but has the capacity to develop a richer, more complete, and more useful understanding of the meaning of things. The conventional approach assigns ordinary meaning to concrete circumstances, whereas functional analysis looks for special meaning depending on the context. The aim of making valid comparisons of interactions across policy processes cannot be achieved if only locally applicable, conventional language is employed. Comparisons must use the same basic terms, regardless of the context or local usage. The eight value categories form part of this stable frame of reference, independent of the society or context, which you can use to make comparisons. If the value component or any other element is omitted from the functional approach, your resulting analysis and insight will be greatly diminished.

This distinction between conventional and functional perspectives comes from anthropology, and professional policy analysts' understanding of human interaction is similar to the work of anthropologists. The functional approach can be further understood by examining how anthropologists classify cultural ceremonies. Lasswell and McDougal (1992) note that in the nineteenth century, for instance, anthropologists used formalistic methods to examine the content of the origin myths of all cultures and classified them by the similarities in their apparent content. Stories that featured the sun were grouped together, for example, whereas stories about the moon were in a different group. We might compare this to the way early taxonomists grouped all red flowers together, assuming that they were closely related. This structural grouping proved faulty. Later on, functional anthropologists, like Linnaeus in biology, studied the entire situation or context to find the role or function that the myth actually played in the society under study. Doing so required much study but paid off in practical results because researchers were then able to comprehend what they were seeing and how meaning was created for the society. Viewing human interactions in functional terms is essential for understanding policy processes and finding ways to improve their outcomes.

One way to distinguish between conventional and functional perspectives is through an example that focuses on the eight value categories. If a mining company loses a court battle and is required to pay a fine for environmental damages, the company clearly suffers a deprivation in wealth. This may be all that a conventional observer sees. For someone who views policy matters more functionally, however, the guilty verdict is also a condemnation in ethical or moral terms. The company's opponents may publicly deride its loss. Its poor record may alienate stockholders and the public or sever access to inside sources of markets and information. The company's operations and practices, and thus the skills, competence, and power of the company's managers, may be challenged. The environmental groups that filed the suit may grow in public stature. These examples are all "measurements" of enlightenment, power, well-being, respect, rectitude, and the other values. Numerous shifts in value holdings—more specifically, in problem definitions, institutional structures, and power of various groups—may result from the court's decision. There is likely to be a complex web of repercussions, implications, and shifts for social process that can be discovered simply by looking at value dynamics—phenomena that tend to be invisible to conventional observers. Because of this, the functional approach offers much greater insights into social process and greater options for solving complex problems.

Convention encourages partial blindness about the policy process. Attention is often turned to the rules and away from the process so that a clear, empirically based picture is not possible. Often key participants are overlooked, the functioning of the decision process is not clear, and social process events are not understood in value and institutional terms, so that the observer is left with, at best, an incomplete, impressionistic understanding or, at worst, confusion. Conventional perspectives are able to describe past events or trends only anecdotally, an inadequate approach for our purposes of proficient, rigorous, practical professionalism.

Confronting the conventional perspective with the functional one is a significant contribution of science to policy (Lasswell and McDougal 1992). Because many participants in a policy process will probably be conventional participants, presenting a functional analysis to them can open up much insight and many opportunities for improvement. You must be sensitive to the problem of language and try to communicate in ways appropriate to the context or circumstances.

Conclusions

In order to participate most effectively in the policy processes that determine how natural resources are used, professionals must cultivate a policy orientation. This requires that they maintain a special kind of perspective, a unique standpoint, with regard to the decision processes of which they are a part. This standpoint should be focused on helping the entire community to clarify and secure its common interests and on serving the particular needs of decision makers. In this sense it is pragmatic because it deals not only with the narrow scientific content of an issue but also with the broader implications of the uses and purposes that scientific knowledge has in decision making and the opportunities it offers. This unique standpoint is also analytical, rational, and integrative; the policy sciences framework provides a thorough guide to both the content and the procedures that professionals should follow. Finally, the standpoint is reflective, requiring professionals to take stock continually of their attention to social process, decision process, problem orientation, problem definition, the value-institution transactions at play, their own roles and perspectives, and a host of other variables. Professionals who are struggling to understand the difficult workings of policy, who have run up against roadblocks in accomplishing their goals, who strive to be "friends of the process," who have tried to avoid the traps of ideology or bureaucracy, or who have begun the transition from a parochial identity to a more universal outlook—all will find the policy sciences framework and unique professional standpoint to be highly beneficial.

SUGGESTIONS FOR FURTHER READING

Schön, D. A. 1983. *The reflective practitioner: How professionals think in action.* New York: Basic.

———. 1987. *Educating the reflective practitioner: Toward a new design for teaching and learning in the professions.* San Francisco: Jossey-Bass.

Sullivan, W. M. 1995. *Work and integrity: The crises and promise of professionalism in America.* New York: Harper Business.

7 Policy Analysis and Multiple Methods

Learning about a particular policy process so that it can be improved and at the same time mobilizing pertinent, substantive knowledge requires diverse methods and skills. But it is not clear that the training that most natural resource professionals receive in the science and management of water, soils, air, forests, range, wildlife, landscaping, or recreation adequately prepares them for policy analytic work. Too often their training creates the misimpression that their work is separate from public policy processes or the power process, indeed, that they should have little to do with policy or values. Whether your work involves examining management problems, preparing briefing papers, preparing management plans, or advising the boss, you are conducting policy analysis, and you need to see yourself as a policy analyst and develop skill in appropriate methods. *Policy analytic work* means carrying out problem-oriented exercises and mapping social and decision processes to improve conservation outcomes. As we have seen in earlier chapters, methods already exist for these purposes, and new ones will no doubt be invented as needed. Professionals generally begin by applying methods they already know, add new ones as they encounter new policy problems (which demand ever more sophisticated analysis), and expand the scope of their methods as they gain experience.

This chapter expands the treatment of the policy analytic process in previous chapters and offers a perspective on the benefits and dangers of policy analysis. It describes the kind of professional preparation needed for policy analytic work and reviews some general methods that are available, methods that permit insight and depth in policy understanding and offer ways to improve policy processes and outcomes.

Policy Analytic Work

Many of the public problems that policy analysts deal with today are "trans-scientific," that is, they are rarely purely technical or purely political. Matters of fact in these problems can be stated in terms of science, but the problems themselves are, in practice and in principle, unanswerable by science alone (Majone 1989). Because policy makers face increasing numbers of and complexity in trans-scientific issues in all branches of government, the business sector, and the nongovernmental community (Lewin and Shakun 1976), they are turning for help to professionals in the social and decisional sciences with diverse methods at hand. This demand has moved analysts to the forefront in natural resource policy and management.

Policy Analysis and Research

The success of policy processes has always been an issue of major concern to the public and, more formally, to social scientists, who have studied the policy-making process in many arenas (for example, Brewer 1973b, 1978; Bardach 1977; Betts 1978; Fesler 1980; Berk and Rossi 1990). A wealth of policy analytic literature has been published, but for newcomers it may be difficult to distinguish among the bewildering array of policy perspectives and methods it contains. To make your introduction easier, I will define a few terms that describe what policy analytic work is, discuss how to start an analytic process, and offer some practical suggestions about problem solving.

There is no one accepted definition of policy analysis, nor is there a single accepted method for its study (Vickers 1965; Caputo 1977). One good definition notes that "policy analysis is an applied sub-field [of the social sciences] whose content cannot be determined by disciplinary boundaries but by whatever appears appropriate to the circumstances of the time and the nature of [the] problem" (Wildavsky 1979, 15). At the heart of all analytical methods is one concern: how to bring appropriate knowledge to bear in policy decisions. On the basis of policy analysis, decision makers bet on the future of their choice and use their limited resources to try to achieve it. In practice, the people who are doing the decision making are seldom the same ones who are generating the policy-relevant knowledge essential for those decisions.

Bringing analysts and decision makers into close contact is a necessary, although not sufficient, aid to policy making. Prospects for improved policy are greater if both share a frame of reference.

Although the terms *analysis* and *research* are often used interchangeably, policy analysis is more popular and loosely used than research. Policy analysis is done in the field, so to speak, whereas research is conducted in more academic settings (Brewer 1973a). Basically, *policy research* is used to study the policy process. It seeks to explain the policy under consideration as well as reasons for its adoption and its probable effects. Most definitions of *policy analysis* have a common theme—client-oriented advice relevant to public decisions (Weimer and Vining 1989). A quick review of some of these adds richness to this meaning. Policy analysis is directed at providing policy makers with the information and judgments they need to solve contemporary problems (Majchrzak 1984, 11), or in Dery's (1984b) terms, to obtain a reliable picture of both the problem and its context. It is examination of public problems in order to clarify issues, alternatives, and consequences and to improve policy decisions (Lynn 1980) in an effort to raise the level of argument among contending participants to produce a higher-quality debate and public choices among better-known alternatives (Hogwood and Gunn 1987). Bednarz and Wood (1991) define it as a formal, research-based means to gather and assess information about an issue requiring a decision. Dunn (1981) echoes these thoughts in calling it an applied discipline that uses multiple methods of inquiry and argument to resolve policy problems. It is thus, according to Rossi and Freeman (1993), a mix of science, craft, and art, and according to Brewer (1981), an exciting and important activity with multiple purposes and uses.

One key to good policy analysis is for you to adopt an overall "professional mindset," such as the policy sciences standpoint described in Chapter 6, rather than merely master a few technical skills such as cost benefit analysis or other rational choice methods. Successful policy analysts are trained to deal effectively with values; the method helps decision makers decide who gets what, when, how, and why—clearly one of the primary functions of social process (McDougal et al. 1988). Analysis is thus not a "cookbook" activity (Beveridge 1950). You must always keep this fact in mind.

Starting the Analytic Process

A number of researchers, among them Weimer and Vining (1989), Quade (1975), Hogwood and Gunn (1986), and Parsons (1995), offer practical advice for beginning the analytic process and "operationaliz-ing" your policy sciences inquiry.

First, policy analysis is never very precise or certain at the outset. It begins when someone—you, your boss, or someone outside your organization—perceives that a problem exists. Weimer and Vining (180) note that "good analysis asks the right questions and creatively but logically answers them. The approach that you choose should allow you to eliminate, minimize, or at least mitigate your particular weak-nesses in thinking and writing." They recommend two primary analyti-cal rules—first, that linear and nonlinear thinkers should adopt each other's styles in thinking and writing, and second, that all analysts should use a rigorous combination of linear and nonlinear approaches. Linear thinkers solve problems by moving sequentially through a series of steps or operations, whereas nonlinear thinkers solve problems "configuratively" by moving back and forth among the steps, data, and intuition.

Weimer and Vining also note that all policy analysis is client-oriented or -driven (clients, again, being whoever is served by your analysis, including bosses, decision makers, colleagues, the public). You should be sensitive to the many professional and ethical concerns involved. They suggest, first, that the analyst must address the problem that the client poses. It is likely that the problem and its solution may involve some uncertainties as defined by the client, and it is best to note ambiguities and uncertainties rather than sweep them under the rug and hope they will go away. It is better to answer the client's ques-tions and in the process reveal the uncertainty than to take some other course of action. Analysts and clients like things to be tidy, but analysis seldom ever is. The analyst must never let the client be blind-sided by events. The client may not appreciate the ambiguities, so it is very important that the analyst fully explain the analysis and recommenda-tions. The second situation that is likely to occur in client-driven analy-sis is that the client's first account of the problem turns out to be the wrong problem (Dession and Lasswell 1955). Weimar and Vining sug-gest that the analyst address this possibility with the client as soon as it is discovered. Presenting a new problem definition may be difficult since the client may be adamant that the original definition was cor-rect. Nevertheless, you and your client should discuss the issue fully. Third, clients often provide problems that are poorly formulated, even if they are the right ones. Clients may only be able to describe the symp-toms, telling the analyst that "something is wrong, but I can't pin it down" or "things just don't seem right," so that it falls to the analyst to discern the core problem. You should carefully and delicately guide the client to a clear sense of the problem. This kind of interpersonal interaction is central to successful policy analytic work.

Edward Quade (1975), who was closely associated with Harold Lasswell and the early decades of the policy sciences, also presents a useful guide to policy analysis. He recommends a two-part process—defining the problem and preparing an issue paper. In order to define the problem, he suggests, you, the analyst, should interview in detail the decision maker or person who asked for the analysis as well as other people associated with the problem. He lists basic questions to ask at this point (67): (1) How did the problem arise? Why is it a problem? (2) Who are the people who believe it to be a problem? (3) If it involves implementing a decision made higher up, what is the chain of argument leading to that decision? (4) Why is a solution important? If an analysis is made, what will be done with it? Will anybody be able to act on the recommendations? (5) What should a solution look like? What sort of solution is acceptable? (6) Is it the right problem anyway? Might it not be just a manifestation or a symptom of a much larger or deeper problem? Would it be better to tackle this larger problem, if there is one? (7) Since analytic resources are always limited, does it seem at this stage that the return from the study effort will be justified, or would this analytic effort be better applied elsewhere? Many related questions should come to mind as you get into the analysis. These questions should allow you to form a picture of the nature of the problem and what to do about it.

Quade indicates that in preparing an issue paper the analyst should keep in mind that it is nothing more than a formal description of the problem definition. It should be a complete assessment of all that is known about the problem, including data presentations, in sufficient detail so that decision makers can act. The issue paper, which should stand alone, should answer such questions as the following (69): (1) What is the magnitude of the problem? How widespread is it now? How large is it likely to be in future years? (2) Toward what public objective should programs for meeting the problem be directed? How can estimates of progress toward the objectives be made? What proxies might be used if the estimates do not seem to be directly measurable? (3) What specific activities relevant to the problem is the government currently undertaking? What alternative programs or activities should be considered?

Quade also suggests several other considerations. First, carefully select the staff to do the analysis. Teams are usually best. Unfortunately, most organizations do not have even one professional policy analyst, much less one or more teams. As a result, it may be necessary to hire a consultant who is both knowledgeable in the policy sciences (or some other tradition in the policy movement) and experienced with the subject matter. Second, to facilitate the analytical task, it is helpful to prepare a work statement specifying what is to be accomplished, estimated times, and personnel assignments. Keep in mind, though, that policy analysis is largely subjective and depends directly on the experience, intuition, and judgment of the analyst. A work statement should not overspecify or constrain the analyst's work. Third, in choosing an approach to the analysis, keep in mind that as it gets under way the problem may appear different than it did at first. Problems seldom remain static, and the "viewing angle" of both analyst and client will change as well. Quade reminds analysts to appreciate that throughout the exercise their job is to "advise a decision-maker, help answer his questions, sharpen his intuition, and broaden his basis for judgment. In practically no case should we expect to 'prove' to the decision-maker that a particular course of action is uniquely best" (79).

Finally, policy analysts are often required to present policy makers with briefs as the final product of their work. Policy briefs are very short (usually one-page) synopses of the problem, the recommended alternative, and perhaps some justification. They should be problem-oriented, contextual, and limited to key points with brief explanations or qualifying statements to support each point. Keep it simple and succinct but realistic.

Benefits and Cautions

The growing role of policy analysis in public deliberation is a reflection of broad changes underway in the role and character of government (see Meltsner 1976; Cortner and Moote 1998). You must be aware of these larger social changes, which are moving you as professionals and policy analysts into the forefront in natural resource work. Analysis plays a basic role in society because many people—from politicians, educators, and business leaders to ordinary people—believe there is a close relation between "what we know (knowledge) and what we do (action)" (Bednarz and Wood 1991, 17). At its most elemental and comprehensive, policy analysis is focused on producing or assessing knowledge that can be used in decisions and actions. Lawrence Lynn (1980, 5) notes that "policy analysis can indeed improve the bases on which public policy decisions are made. . . . To 'improve' in this sense means to make complex issues more intelligible, the range of alternatives more appropriate, and the social consequences of each alternative more evident than they would be if such skills were not used."

Despite the clear benefits to decision makers and the public, there are limitations as well, and you and other would-be analysts should be cautious in your professional practice (see Nakamura and Smallwood 1980; Healy and Ascher 1995).

There are several practical issues you must keep in mind in doing analysis using the policy sciences approach. First, persuasion in the form of words and arguments is central. As all segments of society participate in continuous policy processes—the executive branch of government, the courts, legislative bodies, political parties, the media, special interest groups, scientists, the electorate, and independent experts of all kinds (Pressman and Wildvasky 1979; Latour 1987)—each attempts to persuade others. In your policy analytic role, you need to become highly skilled in the art of persuasion. In some societies, though, persuasion is replaced by coercion in the search for consensus, and in the worst cases, an open, fair process is not permitted and democracy is destroyed.

Second, knowing the social process within which analysis takes place is vitally important to both analysts and citizens. But mere understanding of decision-making contexts is not enough. You must remember that real policy problems are typically messy, ill defined, and resistant to the straightforward application of formal analytic techniques (Lee 1993). Good analysis and methods require cleverness and imagination more than mastery of technique. Policy analysts should be creating a useful product from complex, dynamic social situations, using data, various analytic tools, common sense, and nonquantitative evidence. "A good analyst will never be perceived as a 'hammer looking for a nail'" (Lynn 1980, 6). In other words, don't become method-bound; stay open and flexible and constantly maintain a learning mode.

Third, certain applications of policy analysis are not without critics. You should be on guard for these criticisms and seek to avoid situations in which they are made. Laurence Tribe (1973) of Harvard University Law School, for example, views policy analysis as bad when it reflects thinly disguised monetary, self-interested demands. Analysis is at its worst when it uses discipline-based, analytic methods that are biased in favor of narrow models. Tribe and others believe that public policy should be based on principles to which we are committed or obligated rather than on financial calculations of who gains and who loses by specific government actions. Good policy processes should help us discover and integrate our values, not blindly apply analytic methods built on short-term self-interest or technical considerations.

Fourth, many policy-making approaches are weak because analysis and decision making occur without well-defined operational principles, such as an interdisciplinary focus, attention to context, or acknowledgment of ideological or other biases. This is the case today in many natural resource arenas. The largest single problem is the disciplinary and bureaucratic blinders of participants (Dror 1971a, b); scientists trained in a single discipline and decision makers working for parochial bureaucracies think and act in narrow, self-interested, political terms. Their "operating theories" do not fit real problems and do little to secure the common interest. Many policy analytic approaches are also weak because they are full of the analyst's value judgments, they overlook key variables concerning the problem, they are politically infeasible, or they fail to clarify the criteria for evaluating the success of recommended actions.

Fifth, policy analysis may be either qualitative or quantitative or a combination. Both are scientific, although scientific positivists may disagree. Some people feel that the social and some biological sciences cannot be scientific unless they produce exact quantitative measurements, whether they are appropriate for understanding and resolving problems or not. But a prerequisite to quantitative study and perhaps other more selective research is broad, yet focused, qualitative inquiry. Qualitative research has been and remains essential to policy research (see Lasswell and Kaplan 1950). Although social variables or indices of social behavior change—for example, "justice for all" has had dramatically different meanings in the United States over the past two hundred years, as has "nature conservation"—this instability highlights the importance of trend research and the need to update indices in response to social trends.

There are other weaknesses in much of policy analysis as it is generally practiced. The major problem is that many of those who call themselves policy analysts carry out their activities from a single disciplinary perspective and not from the broad-based, integrative approach encouraged by the policy sciences. Brunner (1991a, 1996a) notes that this problem is widespread in the overall policy movement and that disciplinary-based differences in outlook and method persist, compounding the difficulty. As long as these partial approaches to policy process and analysis exist, it is necessary, he argues, to restrict the term *policy sciences* to the configurative conception of Harold Lasswell and his collaborators, which is the oldest and most comprehensive approach in the movement. You should be aware of this distinction in using policy analysis, because many people who call themselves policy analysts are not policy scientists and because they likely have allegiances to particular disciplines, organizations, and ideologies that narrow their scope and skill.

The Social Planetarium

The ultimate goal of policy analysis is to bring about a situation in which people fully and sensitively come to understand and practically address pressing community problems. In keeping with this goal, in 1938 Harold Lasswell envisioned a "social planetarium" that would provide a comprehensive panorama of the workings of society, just as its astronomical counterpart affords views of the heavens. It would provide a total environment for an observer located anywhere on earth, so that, as the social dynamic moves, so does the viewing angle of observers and the pictures they observe of the relations among observed phenomena. The idea behind the planetarium experience is that it would bring many isolated items into a picture of the whole. Its principal feature is its contextuality. Lasswell felt that applying this notion to the political complexity of our times could bring confused and uncertain groups of people into a shared experience of the past and potential futures of society. If we could create a social planetarium, it might well be the experience to bring sanity and consent to public policy and much of social life. Muth and Bolland (1981) saw Lasswell's imaginative method to study future states of society as a means to revitalize civic order.

The social planetarium idea rests on the fact that all human choices are about the future, yet most methods of problem solving are deficient in presenting the future to us in ways that contribute to insight and understanding. Current approaches are often insufficiently contextual, they do not foster clear perception, they lack clarity in detailing a path from the present to the future, and alternatives are unequally addressed in deliberations about problem solving, especially at large scales—to mention only a few weaknesses. The social planetarium was advanced to address these shortcomings; it could bring vivid perceptions to viewers because it stresses the "looking response." Planetarium presentations could pass through time carrying people through a stepwise process. As Lasswell (1959, 106–7) noted, "existing perceptual images can be challenged and modified in such a way that the novel future possibilities can be grasped. Participants are liberated from the perceptual caves of the present." The planetarium method could be used to strengthen every aspect of the decision process in specific applications and to champion human dignity. Other institutions in society, such as museums of art, local history, or natural history, could all be integrated into the idea of a social planetarium to make applications vivid, real, and practically useful.

Professional Preparation

Policy analysts work in organizations of all kinds, including the federal government, state, county, and municipal governments, business, think tanks, universities, and nongovernmental organizations. But their job descriptions seldom mention policy analysis; in natural resource fields, for instance, they may read, instead, "wildlife planner," "public lands coordinator," or "program manager." All these jobs and many others require policy analytic skills, whether the job advertisement indicates this fact or not. The basic conceptual tools of policy orientation, social and decision process mapping, and standpoint clarification will serve you well in most situations in preparing for policy analytic work (see Gilovich 1991; Halpern 1996). There is, however, no substitute for an organized course or curriculum in which students or practitioners work closely with instructors and professional policy analysts, and there is no substitute for actual, guided work experience to teach the nonacademic, nontechnical aspects (see Clark and Willard 2000). Besides university courses, on-the-job, in-service courses can be organized for this purpose (see the appendix; Clark and Ashton 1999). University training is an important but not a sufficient condition for achieving professional competence; experience in diverse settings is also vital.

Learning Policy Analysis

Would-be analysts must keep in mind that every policy decision is part of a psychological, social, and political context that encompasses far more than the nominal or technical issues at hand. The essence of the work is to use analytic skills to strengthen the foundations of decision making. This requires an understanding not only of the logical structure of policy problems, but also of the decision-making process and the factors shaping it. Analysts who neglect this fact tend to produce work that is regarded by decision makers as misguided and irrelevant, even if it is technically competent (Lynn 1980).

Good policy analysis involves formulating and communicating advice. The first step is for you to analyze yourself, a task that was introduced in Chapter 6. Successful analysts know, for instance, whether they are generally linear or nonlinear thinkers (see Weimer and Vining

1989). Both approaches have benefits, and knowing how you and your co-workers or teammates think and work is essential: this difference may cause conflicts, but complementary styles can also be productive.

Analysts need preparation in five areas (Wiemer and Vining 1989). First, you must be skilled at gathering, organizing, and communicating information. Often strict deadlines as well as limited access to key information and people make the job challenging. You must also operate in a consistent and realistic manner given the role of government and the process of public deliberations. Second, you must possess a method, such as the policy sciences framework, for putting problems into context. Third, you need knowledge and skills in order to evaluate critically and confidently various alternative solutions to problems. Economics and statistics are helpful but often do not get at the fundamental value issues in a policy process. The policy sciences, on the other hand, do attend to these issues and in fact may invoke economics and statistics along with a host of other methods to understand and solve policy problems. Fourth, you must understand organizational systems and their behavior as well as policy processes. Finally, you will face ethical dilemmas as well as analytic challenges in your career, and the ethical commitment required of a policy scientist should be the promotion of human dignity and democracy at all times.

Analytic Roles and Observational Methods

Many roles are open to analysts (Dession and Lasswell 1955; Lasswell 1959; Starling 1988). You should be aware of which ones you are using. Within your standpoint as "a friend to the policy process," helping it to work well in support of democracy or human dignity, you might have one of four roles—participant, spectator, interviewer, or collector (Lasswell 1938). You might be a participant like most other people involved in the process. As an interviewer, you might gather data for scientific purposes about the decision and social process. In doing so you must expect to influence others, which gives rise to a complex technical problem of understanding your own impacts on others as potential sources of error in the process. As a spectator, you might be describing what you see without informing people of your identity or purposes. You might, for instance, serve as an "anthropologist" of sorts, living in society as an ordinary member but systematically describing and analyzing that society or its decision-making efforts. You have two basic tasks—to understand the policy process itself and to communicate results in a useful form to decision makers (Lasswell 1971c). As a collector, you might examine records and use them scientifically to understand the process further. A successful policy orientation requires that you remain flexible and adaptive, adopting one or more of these roles depending on the situation and purpose. Improving knowledge of policy processes is the major responsibility of policy-oriented professionals, although channeling knowledge into action is the responsibility of everyone involved.

There are two basic methodological approaches to policy analysis (Lasswell 1938). The intensive approach is prolonged and complex: the analytic observer concentrates on the people and their perspectives over time and in great depth, including possibly detailed personality study. The extensive approach is cursory and simple: the analyst focuses briefly and superficially, as may be done in surveys and interviews. Methods may also be direct or indirect, guided or unguided, subject-aware or subject-unaware. Direct methods bring you into a direct relationship to the event in question, such as questioning a person about a policy issue; indirect methods deal with past events. Your observation is guided when you ask a person to fill out a questionnaire, but it is unguided when you simply observe someone's behavior at a meeting. Subject-aware methods enlist the aid of people in life history reconstruction, for example, but watching them without revealing your intentions or activities is a subject-unaware method. Procedures may differ depending on the purposes of the analysis and the situation, but in choosing policy analytic methods, the most important point to keep in mind is that, as noted earlier, "every detail is affected by interaction with the total context of which it is part" (Lasswell and McDougal 1992, 876). Thus, any conception or method of inquiry that fails to place a problem in the setting of society is inadequate for the task of science, which is to identify and describe conditioning factors, as well as for the other tasks of problem orientation. Acceptable methods include cases (Clark 1986a, b), correlations (Marcus and Fischer 1988), experiments (Little and Hills 1978), and prototypes (Lasswell 1971a), for example. The use of sound methods will more likely lead to in-depth analysis, clear explanations, and practical recommendations.

A major task in policy analysis is deciding how to simplify complex issues and clarify their broad outlines without losing important detail. In all decision phases and at all levels, it is important to identify social process: participants or organizations, perspectives, base and scope values, situations or arenas, strategies, outcomes, and effects (for example, Manheim 1977). Selecting who or what to map and deciding who is involved are primary issues. Effects should be evaluated in terms of value indulgences and deprivations and changes in institutions and values. It is also necessary to look for rigidities and flexibilities in the social and decision processes that affect the diffusion or restriction of innovations and therefore improvements. Your recommendations should be targeted to eliminate the hold of rigid perspectives and practices and to promote flexibility and adaptation in policy systems. More to the point, you should encourage practices that allow all policy systems to become open-ended learning systems. These are the most adaptive kind. When thinking about alternatives, draw up three kinds of maps: opportunity maps, showing where there are openings for making improvements, obstacle maps, showing how to attack and overcome obstacles, including end-runs and other possibilities, and factor maps, identifying levers that can be manipulated to make improvements.

A Case of Policy Analysis: Global Climate Change

There are policy analyses of many natural resource issues (for example, Doughty 1987; Dovers 1996). We will examine one that is an excellent example of a comprehensive, practical analysis directed by the policy sciences concepts described in this book. This analysis by Ronald Brunner focuses on the problem of global climate change, including the issues of ozone depletion, high temperatures, altered moisture patterns, deforestation, and greenhouse effects that have grabbed international attention (Brunner 1991b; Brunner and Klein 1999). Brunner's analysis indicates that the United States is misdefining the policy problem and is therefore unlikely to find an adequate solution. Scientific, economic, and political answers are, for the most part, the only ones being advanced, but these so-called solutions do not meet the criteria of rationality or the test of practical experience. The problem definition around which national policy is being developed is weak and inadequate, according to these criteria. A popular article by Sarewitz and Pielke (2000) makes the same point.

Why is this happening when some of the country's best minds are examining a problem of such magnitude and momentous consequences? Brunner (1991b) argues that powerful economic and political interests in our culture are reflected in and reinforce these weak definitions and that we have become trapped in problem definitions that cannot be justified. Fundamentally, the problem may be that our culture cannot integrate its own science-based technologies for exploiting nature in such a way that they can be controlled by ethical or political guidance, and because of this our culture jeopardizes its own sustainability. The best, first step toward a solution to this cultural problem, suggests Brunner, is reasoned action supported by political leadership that "challenges selected elements of the dominant culture and directs attention to alternatives—thereby opening up the search for solutions" (1991b, 291).

Brunner (1996b) contends that the U.S. Global Change Research Program (USGCRP) is on the wrong track to find policy solutions to global climate change, which is just one possible disruption from unprecedented scales of human activities on the planet. The project, which is the center of the U.S. response, is large—in FY 1993, it incurred $1.3 billion in direct costs. It centers on developing a predictive model of the integrated earth climate system as a basis for rational, comprehensive, and cost-effective policy. That is, a scientific model is being promoted to justify policy. But, as Brunner observes, the program's performance has fallen far short of expectations, and Congress and the public are beginning to note the shortfall. His analysis identifies several foreseeable and preventable weaknesses in USGCRP and recommends remedial actions. Basically, the scientific model overestimates the capabilities of science and the probability of consensus on policy. Offering substantial support for his analysis, Brunner proposes that decentralized policy teams integrate all scientific and nonscientific issues and concerns into information and recommendations easily usable by policy makers. He argues that we need to get away from the big science of past decades and move to smaller-scale, learning-focused, and highly usable approaches in this and other natural resource problems.

Methods for Improving Policy Process

The methods you choose for any particular analysis will depend on which dimensions of a policy process you want to bring into relief. You can't look at the entire policy process all at once. Your analysis may require that you initially focus on social process mapping, decision process appraisal, or problem orientation. An outline for examining a policy problem contextually and recommending alternatives is offered in table 7.1 (see Clark, Willard, and Cromley 2000). Overall, though, you will need appropriate methods to examine all these dimensions of a policy problem and to integrate your knowledge into a picture of the whole (Lasswell 1971a). This is a tall order, but various methods of observation, analysis, and policy improvement are available (for example, Putt and Springer 1989).

Described below are four major analytical methods that are particularly helpful in fostering the kind of problem-oriented, contextual inquiry I have promoted throughout this book. All depend on longer-

Table 7.1. Suggested Outline for a Thorough Paper on a Particular Policy Process

TITLE (less than ten words)
Your name, affiliation, address

ABSTRACT (less than 200 words)

INTRODUCTION
A. Be problem oriented, give goals, problems, alternatives (e.g., "The policy problem is . . .")
B. List the purposes of the paper
C. Clarify your standpoint in reference to the problem
D. Describe methods you used

PROBLEM (description of the policy problem)
A. Specify contextually (e.g., social process) and in some detail the problem that is the subject of your study
B. Specify problem in terms of decision process
C. Clarify goals in reference to the problem of concern

ANALYSIS OF THE POLICY PROBLEM (trends, conditions, projections)
A. Describe trends in the decision-making process that have had an impact on the problem of concern; identify particular impacts and their relation to the achievement of goals
B. Identify and examine the factors that have shaped those trends and impacts
C. Project future trends in decision making and accompanying impacts; emphasize the relationship between projected impacts and the achievement of goals

RECOMMENDATIONS (alternatives and justification)
A. Present alternatives for resolving the problem, given the projections and conditions described above
B. Evaluate the alternative strategies you proposed for their potential contribution toward reaching the goals
C. Select and justify particular strategies to resolve the problem

CONCLUSION
A. Very briefly state the goals and problems
B. Recommend your solution and justify it

ACKNOWLEDGMENTS

LITERATURE CITED

Note: Shorter briefing papers should include as a minimum one paragraph describing the goals, the policy problem in relation to the goals, and recommendations (a list of alternatives and selection and justification of one or more). Very brief papers should give the conclusions of your analysis. See Clark and Willard 2000.

term study, data and input from multiple people and sources, careful, in-depth analysis, and exploration of multiple scenarios, models, opportunities, and alternatives. They all promote creativity, innovation, adaptability, and learning.

Case Studies

Policy analysis can be presented to your clients in the form of a case study for easy reading and comprehension. Common in policy analysis, case studies are systematic records of a process with the objective of learning and communicating (Taylor 1983; Clark 1986b). They place boundaries around a particular location, time, and set of events, describing real situations in detail and illustrating complexities and subtleties of policy process. There are four types of case studies, distinguished by their complexity. First is the technical, problem-solving case, which is usually short, fact-laden, and well ordered. A single "best" solution is sought, based on a specified analytic model such as cost-benefit analysis. Second is a short, structured vignette, which lacks one "best" solution but lends itself to a "better" solution given the conceptual problem-solving framework provided. Third is the long, unstructured problem, or "opportunity-identifying" case. Real-life complexity is introduced and the exact problem and its underlying causes are not clear, so that the "best" or "better" solution is in question. Adequate theory and precedent are provided, however, to allow a preferred conclusion to be derived. Fourth is the ground-breaking case that includes new situations in which little if any direct information from previous systematic policy research is available. Only relevant theory is available to address the case. There is abundant material to guide preparation of case studies (for example, Golembiewski and White 1983; Yin 1989; Edge 1991).

There are many benefits to the case study approach. It allows researchers to discover how decision processes work and the effects of decisions over time, and it allows decision makers (or students) to examine possible solutions without costly real-life consequences and to profit from the combined analysis and judgment of analysts. Case studies can thus serve as models for sharpening responsible judgment. Lynn (1980) argues that cases permit analysts, students, or others to gain practice in defining or conceptualizing policy problems in order aid decision makers. They also help people learn how to choose analytic approaches and design new ones, identify policy-relevant information, construct successful policy arguments, and communicate results and recommendations in a style appropriate to real decision-making contexts. In addition, case analysis and presentation force analysts to appreciate the time constraints on officials and the need to make decisions whether their understanding of the problem is complete or not. They can also highlight the features of ill-defined policy problems and the fact that policy decisions are embedded in larger political and bureaucratic contexts.

The use of cases in decision making has pitfalls, however. Analysts may be encouraged toward superficiality because they are often forced to make decisions and recommendations after relatively few hours of study (Lynn 1980). They may also acquire technical and analytic skills without recognizing that their ultimate use is in the service of political processes or without believing that analytic sophistication has any bearing on effectiveness in the political world.

Prototyping

Another method of acquiring knowledge about a policy system is prototyping, which is a small-scale trial intervention in a policy or social system with the express intent of learning about the system and improving outcomes. There are many advantages to such demonstration projects, not the least of which is that they are a practice-based, hands-on approach that requires people to work in the field.

Prototypes are innovative approaches to problems, developed on a small scale but geared toward development of a model on which to base future actions or programs (Lasswell 1963). They precede "pilot projects," which are trials of what is considered to be a solution to a problem. As a means to upgrade professional and organizational practice and knowledge in general, prototyping is a constructive response to the need for innovation and creativity. Successful prototypes (or even selected elements of such efforts) encourage other people to adopt them, thus providing a model for replication and continual revision (Lasswell 1963). Successful models can be copied via pilot projects or full-scale adoptions and incorporated into ongoing social and decision processes.

Among the most prominent and successful examples of prototyping was the Vicos project in Peru (Dobyns et al. 1971), in which American anthropologists and policy scientists intervened in a peasant community to bring about increased human dignity. The Americans bought a hacienda from a Peruvian bank, which had acquired the land from an insolvent landowner. Along with the land and buildings came several hundred peasants who had worked the land for generations. With the full cooperation of the peasants, researchers helped the community rise from powerlessness and destitution to a self-governing, relatively prosperous condition in a few short years. Eventually the peasants purchased the land for a small sum and became the collective owners. The Vicos model is a good one for developing communities that are seeking human rights and sustainability. Prototypes are also commonly employed in the business world; auto manufacturers, for example, set up prototypes of varying kinds, ranging from special problem-solving teams to experimental car designs (Westrum 1994).

Prototypes differ from planned pilot studies in that they remain more flexible and creative. Even though the goals of prototyping are clear, numerous ambiguities usually remain. Like case studies and unlike experiments, they cannot be tightly controlled, although they can be partly replicated. Since the aim of any prototypic study, whether official or unofficial, is to devise better strategic programs, it requires clear, detailed, and comprehensive explanations, including all actions undertaken (Lasswell 1971a). The process should stay in the hands of those who are dedicated to increasing knowledge and professional skill, rather than those who are interested in power.

Prototypes facilitate self-observation, build insight, and enhance prospects for success. They are particularly useful for conservation programs, which cannot be treated simply as scientific problems (because their conditions cannot be tightly controlled in a scientific sense) or solely as political problems to be managed only by bureaucratic officials or politicians. Although it has been used successfully in endangered species recovery (Clark et al. 1995), the prototyping strategy has not been applied widely, explicitly, or systematically in natural resource conservation. Where such trial interventions in social systems (and appraisal of their results) have occurred (for example, Miller 1996), they have usually been unconnected to prototyping theory, which could have made the work of comparison and appraisal more efficient.

A number of books have examined diverse initiatives around the world to find successful models for the management of natural resources. They can help those of us interested in prototyping compare "best practices" and learn how to upgrade our work. For example, White et al. (1994) compare prototypical programs of collaborative and community-based management of coral reefs and draw valuable lessons from these experiences. Friedmann and Rangan (1993) compare the diverse environmental movements in Africa, Asia, and Latin America, treating them as exemplars from which to learn how to integrate the needs of local people with complex contexts of locality, environment, and culture. Pye-Smith et al. (1994) examine ten successful cases of communities seeking sustainable use of their environments. Miller (1996) compares nine bioregional or ecosystem management initiatives from around the world, including the Wadden Sea, the Greater Serengeti ecosystem, Great Barrier Reef Marine Park, and the Hill Resources Management Program in India. He sought to learn what worked and what did not and on this basis offers guidelines for better bioregional management, focusing on building capacity (knowledge and skill), stakeholder participation, and cooperation among institutions at appropriate scales.

Keeping a few considerations in mind can help you invent and implement prototyping projects more successfully. First, all people involved in the program should agree to participate, although not everyone need fully understand the exercise. Second, leaders should agree to the general principles and approach of prototyping. Third, the process must be open and creative. Fourth, top professionals should be included and their opinions respected. Finally, the people involved should be interested in improving performance rather than power (that is, keeping politics to a minimum) (Lasswell 1971a). Work settings that are characterized by high levels of complexity, uncertainty, and conflict benefit most from this type of intervention (Brewer and deLeon 1983), but these same features often go hand in hand with issues of power and control, which work against prototyping. One situation in which prototyping could help is the design and operation of species recovery teams, reintroduction efforts, and appraisals of conservation progress, but individuals and interests who prefer the status quo may strongly oppose new ways of interacting, and for prototyping to be effective, participants must neutralize such opposition.

Policy Exercises

Policy exercises are various methods for integrating information that is to be used for policy making, information being the key resource and basic means by which all human problems are appreciated and addressed. Lasswell originated the idea of policy exercises in a series of unpublished memos to his colleagues in the 1970s. He noted that the information we collect for decision making strongly influences how people live, now and into the future, but that numerous problems impede our use of knowledge. Although we now have vastly improved means of scientific observation and measurement, our ability to analyze, interpret, and use data for social and policy ends has not kept pace with the environmental and social problems we face.

Garry Brewer (1986), a student of Lasswell's, attributed the difficulties of creating and using policy-relevant knowledge to the complexity of the systems involved, limited theories, weak methodological tools, and disciplinary searches for a single optimal solution, among others. Moreover, he claimed, the world is full of surprises, and people show both wishful and fearful thinking. For all these reasons, perspectives on both problems and solutions vary and clash, and information is not used well. The way knowledge is produced and used in many societies is a product of positivism and self-interest, but this approach often limits creative efforts, supports special interests rather than common interests, and denies the legitimacy of different perspectives and preferences by adhering too narrowly to limited conceptions of what constitutes reliable knowledge. He recommended a series of new, experimental "policy exercises" to integrate the needed knowledge.

Policy exercises are explicit procedures in which goals are systematically clarified and strategic alternatives created and evaluated in terms of the values involved. Their purpose is to "release for the common good the creative energies of ourselves and others" (Brewer 1986, 467). They include large amounts of scientific data as well as less positivistic data on, for example, people and their perceptions. Explicit policy exercises could be particularly useful in developing policy for integrating knowledge and improving insight, judgment, and action in the management of ecosystems. For example, the Greater Yellowstone ecosystem is ripe for a policy exercise to identify key ecological and sociopolitical trends (see Clark et al. 1999). A wealth of information has been collected by numerous organizations about the region's natural resources, communities' economies, and other factors, but it has not been synthesized or organized in a way to inform decision making (both constitutive and ordinary) in the face of many onrushing problems in resource management. Policy exercises could also be used to clarify management goals, define policy and management problems, sort out conditions under which trends in the ecosystem have taken place, project what is likely to happen in the future, and begin to solve the problems that face the region.

Modeling is one method that can be used in policy exercises. A model can be thought of as a "representation of an entity or situation by something else having the relevant features or properties of the original" (Brewer 1986, 458). Many models have simplifying assumptions and limited application. Analytic models are widely used, for example, in game theory, but are usually quite abstract with few variables explicitly considered. Numerically defined variables must be well understood and their unit of measurement precisely known. Models can be used to explore, plan, crosscheck, forecast, find group opinion, and support advocacy. Evidence indicates that large-scale modeling, however, is not meeting its promise to inform decision and policy process for reasons that are deeply embedded in the perspectives of the people who commission, build, and use them. Brewer noted that large computer models are more often an obstacle than an aid to creative, exploratory thinking about a problem and its solution. It is seldom appreciated that models are undertaken not for scientific purposes but for policy analysis and that no matter what their construction, models embody a single view of a problem, a problem that is sensitive to and qualified by the steps taken to formulate it. Also, judgments and interpretations are involved in deriving meaning from models, although having different groups work on different models of the same problem can overcome some of these difficulties. Examples of models that have aided management of the Yellowstone region to date include Boyce and Anderson's (1999) computer examination of the role of predators at the ecosystem level and Singer and Mack's (1999) exploration of the relation of both fire and carnivore predation to ungulates. Simulation is another method in which a system or organism is represented by another, simpler, more abstract system or model. Digital as well as and analog equipment is used. Many natural resource policy issues use simulation modeling.

Games offer promise in meeting many natural resource management and policy challenges. Methods such as free-form manual gaming involve teams and a referee group operating within the context of a scenario. Computational equipment may be used, but results are not scientific. Games are designed to characterize plausible future situations, and usually positions, expectations, perceptions, facts, and procedures are all challenged. Free-form manual games seek to foster creative thinking about very complex physical and social phenomena— the kinds of problems that seldom capture the attention of the public or its leaders for very long (for example, the loss of biodiversity or global climate change). They seek to identify interventions that may avert unwanted outcomes or to obtain benefits otherwise lost.

Scenarios are the fundamental building block in all modeling and analysis. They form the very basis for bounding and structuring a model and contain the criteria to evaluate it. Because they rely on verbal descriptions, they are widely accessible to many people. Scenarios use many methods, including analytic models, machine simulations, man-machine methods, free-form games, seminars, and group studies. Scenarios are an inexpensive way to meet challenges, although they vary in quality and in their ability to fulfill the game or research purposes for which they were intended.

Another kind of policy exercise is advocacy, or promotion of a perspective, disciplinary paradigm, or a policy preference. Making a case for something is a reasonable intellectual and moral procedure, but, of course, fairness should be observed so that all advocates have a say. In the courts and in arbitrated proceedings, promotion must be separated from personal condemnation for the loser. The key is to make the best case you can and let the public or officials make a determination.

These methods may be learned in formal settings, such as universities or in-house workshops, or in actual problem-solving settings, such as ecosystem management in Greater Yellowstone or in a sustainable development project anywhere in the world. Brewer (1986) has detailed these methods, offered additional references, and given examples of their successful application. You are encouraged to go to these sources and learn the methods they describe.

Decision Seminars

One powerful design that you can use to improve policy is the decision seminar, a group problem-solving method originated by Lasswell (1960b, 1963, 1966a,b, 1971a) that requires creativity, skill, and craft. Decision seminars are designed to encourage specialists and decision makers to integrate their knowledge to address complex problems (Brewer 1975). Key ingredients include substantive knowledge, flexibility, an ability to abstract, and a willingness to construct and rebuild as many representations of a problematic situation as imagination and time permit. Problem-oriented and contextual, they hold much promise for improving natural resource management and policy by employing strategies to manage diverse data sets, sharpen insights into the problem at hand, explore possible solutions, assign institutional responsibility, and appraise outcomes and effects. Brewer (1975, 1986), Burgess and Slonaker (1975), Bolland and Muth (1984), Muth (1987), Willard and Norchi (1993), Clark (1997a), and others have used them in diverse natural resource settings and in other arenas (for example, public education). You can review these studies to learn how decision seminars work and discover ways to incorporate them in your professional work.

A decision seminar is a problem identification and solution exercise that may be a prelude to an official decision process. Members actively, consciously, and systematically carry out problem orientation, social and decision process mapping, and standpoint clarification using multiple methods. They are empirical as well as normative (value-oriented). A seminar's membership is usually limited to ten to fifteen highly talented and variously trained people selected for their unique skills, perspectives, and concerns regarding the problem at hand. Philosophers, historians, scientists, psychologists, those with literary and artistic skills, and individuals in other fields must seek to comprehend and appreciate legitimate differences among themselves. All members must genuinely engage themselves in the work of the seminar. Continuity and commitment over the lifetime of the seminar are absolutely essential, although from time to time, depending on the issue under discussion, outside participation is encouraged. All members should be on guard against the seminar devolving into a conventional conference or seminar format. Even though the core analytic team should be fully aware that its deliberations are nonbinding and unofficial, decision makers may be involved. The team should work consensually: seminar members must overcome the grip of conventional identifications with policy positions, disciplinary paradigms, or organizational loyalties. Mutual respect is an essential ingredient although it is hard to achieve and maintain without considerable work. Judgment is another vital ingredient. The bases for judgments should be made clear even if they are intuitive, since biases and assumptions usually underlie judgments, and knowing what these are, who is promoting them and why, and what their consequences might be can help the group in many ways.

Decision seminars use many different methods as they proceed through complex, value-laden exercises. Members may need to invent new techniques to address the problem at hand. Tools commonly used in decision seminars are maps, drawings, graphs, and pictures, both static and dynamic. Verbal tools are required. Persuasion, discussion, and other forms of constructive engagement are all necessary. Computer teleconferencing can aid a decision seminar. If models are used, the advice is to "model simple and think complex." Methods should be purposively eclectic. The point is to make as many renditions of the complex setting and problem as needed. Critical imagination is key to a successful decision seminar.

Other Methods

There are many additional methods that you can use to orient yourself to a problem thoroughly. You should select ones that will help you in your efforts to improve the substantive and procedural rationality in problem solving (Simon 1996). Traditional, positivistic science can aid substantive rationality, whereas methods in the applied sciences, such as operations research, artificial intelligence, and expert systems can help you improve procedural rationality. A number of reviews and handbooks, specifically for research methods in the social sciences, can help policy-oriented professionals learn new methods and outlooks (see Dominowski 1980; Barzun and Graff 1985; Miller 1991; Dey 1993; Rosaldo 1993; Denzin and Lincoln 1994; Strauss and Corbin 1994; Isaac and Michael 1995). Two particularly useful methods are described below.

Q-methodology has been used to facilitate public involvement in natural resource management (Steelman 1997), enhancing the role of the public and diminishing that of the technical policy expert. It has been applied to ecosystem management of the Chattanooga Watershed, straddling North and South Carolina and Georgia (Maguire 1995), and the Monongahela National Forest Service planning process in West Virginia (Steelman 1997), for instance, and could be used more widely in natural resource policy situations.

The Q-method was developed in the 1930s by William Stephenson as a practical method to facilitate the scientific study of human subjectivity (Stephenson 1964; Brown 1980; McKeown and Thomas 1988; Durning 1996). It is a discourse-based approach to policy analysis that overcomes many of the limitations of trying to find policy solutions through positivism. It is inductive, yet systematic, providing data on the subjective status of public viewpoints. It entails asking people involved in an issue to rank a set of statements that they themselves have made—not facts but opinions about the issue in question—according to a scale. This approach presumes that the meaning of each individual's response depends on the context; it thus asks respondents to consider how their preferences relate to each other. Factor analysis of the ranking is done, and the factors that emerge indicate the areas or categories of subjectivity that exist. Often a common view emerges from a group, or areas of agreement and disagreement about a policy are identified. In turn, this information can be used to find publicly acceptable management practices.

Another second useful approach consists of packages of field methods that arose in the 1990s called rapid rural appraisal (RRA) and participatory rural appraisal (PRA). Each has been applied in communities in developing nations (Somluckrat et al. 1985; Molnar 1989; Bentley 1994; International Institute for Environment and Development 1994; Byers 1996; Freudenberger 1997). These two methods are used to map social and decision processes and orient researchers to problems as seen by local people; the idea is to collect accurate, timely data about a particular situation in order to inform policy debates and to plan and implement local development activities (Freudenberger 1997). Carried out predominantly by researchers, RRA seeks to gather quality information that captures the complexity of local situations and attributes value to local knowledge. A team of outsiders builds rapport and works closely with the local community in gathering data. In contrast, PRA brings the community under study directly into the investigation and seeks to build local capacity to define, analyze, and solve problems. Local people, sometimes aided by outsiders, carry out the basic research. These methods are a combination of applied anthropological, sociological, political science, and natural resource mapping tools.

The clear intent of both RRA and PRA is to conduct problem-oriented, context-sensitive inquiry directed at improving human dignity, and both address many elements of social and decision process, particularly some of the value categories. The data collected, however, would be more complete and rational, presentations to decision makers would be more useful, and better comparisons among communities could be obtained if the policy sciences theory and framework were used systematically and explicitly to guide RRA and PRA.

There are other guides to understanding policy and the analytic process, although few of these are as complete or as sophisticated (neither simplistic nor technique-bound) or offer as high standards as do the policy sciences. It would be best if professionals in natural resource fields kept abreast of developing methods, guides, and theory in policy-related fields. One helpful book is Coughlan and Armour's (1992) *Group Decision-making Techniques for Natural Resource Management Applications,* a guide to avoiding the pitfalls of "group think" as described by Irving Janus (1972; see also Clark et al. 1989). Offering general principles to improve group problem solving, it is problem-oriented, conventional, and easy to read. It promotes the use of multiple methods and evaluates more than two dozen. It would be a useful addition to any policy-oriented professional's library. Hogwood and Gunn (1986), *Policy Analysis for the Real World,* is another down-to-earth text that offers practical methods, and Margoluis and Salafsky (1998), *Measures of Success: Designing, Managing, and Monitoring Conservation and Development Projects,* is a rich source of field methods for planning projects and learning from your experiences. These and many other books can refine and expand your knowledge of the policy aspects of managing natural resources.

Conclusions

Natural resource professionals will be called on throughout their careers to offer policy analytic advice. Their advice will be most sound if it is based on explicit knowledge and skills of policy analysis and research. There is no simple set of steps to be followed in policy analysis, but the analytic framework of the policy sciences is invaluable in guiding analysts to a complete array of variables that should be considered. The precise methods used in any given exercise should be determined by the nature of the problem addressed, characteristics of the context, and the style, judgment, and creativity of the analyst. In any case, good analysis should focus on a given policy context and set of actors, it should be clear about its purposes and its "clients," it should recognize the scope of the problem at hand and thus the need for interdisciplinary thinking, and it should deal with the practical problems of recording, integrating, and interpreting different kinds of data in meaningful ways and communicating results effectively. Better policy analysis and the use of diverse methods will guarantee much better management of our natural resources.

SUGGESTIONS FOR FURTHER READING

Brunner, R. D., C. Colburn, C. Cromley, and R. Klein, eds. 2002. *Governance and natural resources in the American West.* New Haven: Yale University Press.

Coughlan, B. A. K., and C. L. Armour. 1992. Group decision-making techniques for natural resource management applications. Resource Publications 185. Washington, D.C.: U.S. Department of the Interior, Fish and Wildlife Service.

Hogwood, B. W., and L. A. Gunn. 1986. *Policy analysis for the real world.* Oxford: Oxford University Press.

Margoluis, R., and N. Salafsky. 1998. *Measures of success: Designing, managing, and monitoring conservation and development projects.* Washington, D.C.: Island.

8 Natural Resources, Human Rights, and Policy Learning

People everywhere share an interest in a healthy environment. The growing human rights movement is well aware that a quality environment is requisite for a rich, full life and that none of us can thrive and achieve our full potential in a polluted, degraded environment (Zerner 2000). This seems self-evident. To achieve a world of human dignity in healthy environments—sustaining the centuries of labor by untold millions that have brought us to this level—will require us to learn as quickly, as earnestly, and as wisely as possible how to manage our resources and our own behavior. This challenge may be the greatest test before us. Your job as a policy-oriented professional is to foster this collective, intergenerational task.

This chapter offers a perspective on the status and relation of natural resources and human rights, especially as they come together in the idea of sustainable development, and it recommends that policy learning is the key.

Natural Resources and Sustainable Development

For decades national and world leaders have called for better natural resource policy and management. In 1969 U.N. Secretary-General U Thant (1969, 4) detailed the growing "crises of worldwide proportions" that include runaway human populations, poor use of technology, loss of agricultural lands, unplanned urban development, loss of space, and extinction of many plant and animal life forms. In the ensuing decades others have expounded on these global crises in increasingly urgent terms (for example, Ehrlich 1997). In light of these trends, two issues need to be emphasized. First is that we all live in a complex, indivisible web of interrelationships and interdependencies. Second is that the human institutions and practices through which we try to meet our many physical, social, and psychological needs and demands have significant impacts and effects, both beneficial and harmful. Chronic disregard of these considerations at both local and global levels has led to misuse of natural resources and technologies, environmental disasters, and massive value deprivations such as ignorance, poverty, and disease. But during the past few decades consensus has been coalescing about how to address these crises and what might be an appropriate overriding goal for the world community to pursue. This new idea promotes neither rampant exploitation nor strict conservation of natural resources or preservation of all environments in wilderness condition. Instead, it recommends an appropriately conserving yet efficient and positive employment of natural resources for human dignity enjoyed as widely as possible. This seems, ideally at least, to be the philosophy behind *sustainable development.*

Appreciation of the environment and natural resources has undergone dramatic changes as perspectives have changed over time. For example, until recently, environmental health did not figure prominently in policy making; it was a marginal concern distantly related to the more consequential issues of economic growth, human health, and national security. Today, it is widely understood that our present uses of the earth and its natural resources are compromising future use, and the linkages between the environment and our well-being are better known—although still short of being fully assimilated—and are centrally important to many governments, publics, and businesses. In the last few decades of the twentieth century, the ideas of sustainability and sustainable development rose to prominence in research, policy, and political agendas. The global community has begun to recognize the growing disparity between the human rights goal and current realities, and it has begun to consider sustainable development as the preferred alternative, the practicality of which must now be evaluated and compared with other options, including doing little or nothing.

Sustainable development was the major theme at the U.N. Conference on Environment and Development in 1992 (the Rio Conference), a new and urgent agenda to meld ecological concerns with poverty and security concerns (United Nations Environmental Programme 1992a). According to Stephen Dovers (1997), who provides a good review of the modern idea of sustainability, other milestones in its development include works by William Thomas Jr. (1956), Rachael Carson (1963), and the International Union for the Conservation of Nature and Natural Resources (IUCN et al. 1980, 1991). The 1987 World Commission on Environment and Development, or Brundtland Commission, was probably the most influential in defining the modern notion of sustainability and focusing world attention on it.

Numerous authors have described the declining status of the global environment (for example, W. Clark 1986; W. Clark et al. 1990; Kates et al. 1990; World Conservation Monitoring Centre 1992). These works, in turn, have led to calls for better policy and management (for example, United Nations Environmental Programme 1992b; Norse 1993; World Resources Institute 1994). In the United States and elsewhere, this concern is couched in the language of ecosystem management, integrated management, and watershed management, and improved policy is being called for by government, academic, and conservation leaders (for example, U.S. Department of Interior 1993; U.S. Fish and Wildlife Service 1994; Pickett et al. 1997). Internationally, many development projects are underway to bring about sustainability. People from many different arenas are focusing the policy debate on how to move toward human dignity and a sustainable world. Sustainable development is the means preferred by many governments.

Human Rights

Throughout the world, natural resources are being increasingly seen as a common legacy of all of people. International laws and conventions to protect Antarctica, ocean fisheries, space, and global biodiversity illustrate this emerging view, as do the broader criteria used to evaluate practices of exploiting natural resources (in terms of their consequences for all people rather than the economic benefit for a few). For example, cutting of tropical rainforests and the concomitant loss of biodiversity are viewed as harmful to the entire biosphere and to all people. Qualitative assessments of impacts on the global environment and prospects for achieving a genuine commonwealth of human dignity are becoming more comprehensive. The conclusion seems to be that all natural resources and the environment must be protected and maintained as the keystone for improving the quality of human life, indeed, all life. But bringing about new practices so that we can sustainably manage ourselves and our natural resource base is a formidable task (see Clarke and McCool 1985). Many institutions and practices currently threaten the integrity of the biosphere and thus our future, and to achieve practices that preserve both indefinitely will require significant changes in policy processes and value-institutional arrangements in all societies.

Since World War II, demands for human rights everywhere around the world have accelerated dramatically (McDougal et al. 1980; McDougal 1992a), continuing a centuries-long trend and many great historic movements. For example, the English, American, French, Russian, and Chinese revolutions, which were supported by natural law and religious traditions, all originally called for improved human rights. Especially in the past two hundred years, many societies have gradually transformed themselves in the direction of freedom, equality, fairness, and community solidarity and sought to protect themselves from despots, tyrants, dictators, and coercive institutions and activities of government and the private sector. Over the course of history individuals have changed their relation to the state, demanding physical security, inviolability of the person, freedom from cruel and inhumane treatment, freedom from arbitrary arrest and confinement, freedom of conscience and religion, freedom of opinion and expression, and freedom of association and assembly. Demands for civil liberties have become more organized, more insistent, and more universal. It is abundantly clear that the vast majority of people want a system of public order that affords them the widest possible scope for the shaping and sharing of values at all scales. Such authoritative and controlling calls are contained in the constitutions of many countries, including the United States, as well as the Universal Declaration of Human Rights, the International Covenant on Civil and Political Rights, and the International Covenant on Economic, Social, and Cultural Rights (Lasswell and Holmberg 1992; McDougal 1992a).

Trends and Conditions

In the past two centuries, the world has experienced rapid industrialization, massive concentration of wealth, rampant urbanization and suburbanization, depletion of natural resources, drastic environmental alterations, and accelerating rates of change in these and other areas. Many people have benefited from these changes but many others have been harmed by them. In terms of the natural world, there are many unsustainable practices that waste and deplete resources without regard for future generations. Many resources are overexploited for short-term gain or diverted for destructive military purposes. In some cases the raw resources essential for human dignity are lacking, whereas in others the perspectives and institutions needed to use adequate resources sustainably are lacking. One product of the intensified exploitation of natural resources and other people is that huge disparities have developed in wealth, employment, housing, medical care, education, skills, and other important categories.

The conditions that have led to these disparities are complex and dynamic, including a mix of environmental, or external, factors as well as predispositional, or internal, factors. Environmental factors, the most important of which are population, natural resources, and institutional arrangements and practices, figure in every aspect of social process. First, the world population is experiencing explosive growth unprecedented in all of history. This major trend affects all other global components, including the atmosphere, oceans, land masses, and biodiversity. The number of people in the world has profound implications for the sustainability of human rights and natural resources. Population size, consumption patterns, and growth rates are causing major problems worldwide, many of which are likely to worsen in the foreseeable future. In recent decades, the population-resources-technology imbalance has contributed to the deprivation and nonfulfillment of human rights, producing hunger and malnutrition, slums, shantytowns, crowding, disease, stress, deteriorating environments, depletion of finite resources, poverty and poor living conditions, unemployment, illiteracy, crime, political extremism, and a propensity toward violence, to mention only a few problems.

Another external factor is natural resources. If modern scientific research has produced one significant conclusion, it is that all life, including humanity, constitutes a single planetary biosphere—which is currently in jeopardy! McDougal et al. (1980) make the case that a major cause of the failure to meet human rights demands is the mismanagement and misuse of natural resources. It is the principal cause of the worldwide ecological crises that directly threaten the quality of life on the planet. The nature and degree of the crises vary by locale— loss of tropical rainforests and their biodiversity and ocean fisheries depletion are but two examples—but their global ramifications are undeniable. This situation is unprecedented. Yet resource management remains fragmented and inefficient because many barriers hinder the free flow of people, ideas, technology, goods, and services (especially across national borders). Just as natural resources are unevenly distributed over the globe, McDougal and his colleagues note, there are marked discrepancies in patterns of consumption. Some regions contain few forests, poor topsoil, little water, scarce fuels, and few minerals, and thus prospects for growth and more value indulgences in such areas are negligible. Availability of resources is also uneven, a function of technology, manpower, social organization, and other factors. As fresh water, food, energy, and materials are divided among ever growing numbers of people, it seems likely that it will become harder to fulfill human rights demands. Competition will probably intensify and wars of resource distribution may well break out if these conditions are not addressed.

Two more environmental factors responsible for the failure to achieve human rights are institutions and practices. The solutions to many environmental problems (for example, ozone depletion) require integrated worldwide efforts, but at present the institutions and practices needed to tackle global problems are generally too parochial and short-term to be effective at larger scales and for long periods. The accelerating change (including increasing globalization) caused by the universalization of science and technology seems to be outstripping our ability to manage our practices without significant environmental damage. McDougal and his co-authors point out that the world is divided into nation-states and will remain so into the foreseeable future, and elites in each nation play the world social process to their advantage according to the maximization postulate. A strong emphasis on maintaining and promoting sovereignty, nationalism, and rival ideologies makes the task of cooperative environmental management extremely difficult.

Predispositional factors also figure prominently among the conditions that have led to significant discrepancies between the demands for human rights and their attainment. These include the fundamental demands, expectations, and identifications of people, and they are closely associated with weak institutions and practices in many instances. Many predispositions tend to support special interests at the expense of common interests, people's expectations are frequently calculated toward short-term payoffs, and their identifications are often narrow and provincial. Even though many voices actively promote common interest demands such as human rights, the actual patterns of these demands are frequently less than comprehensive—nationalist common interests, rather than those of the world community, for instance. Many people, in fact, are motivated by competing ideologies, contending systems of public order, and divergent forms of parochialism, rather than an overall commitment to human dignity worldwide. Although the combined revolutions of science, technology, and globalization have tended to create broader identifications, and interdependencies among global community members are growing in all value categories, national and subnational identifications remain very powerful, and parochialism continues to flourish in many forms. We all tend to regard other people in terms of economic class, education, job status, race, and so on rather than in terms of our common humanity. At best, our identifications are ambivalent and torn between narrow-mindedness and recognition that all humans, all life forms, are part of a single global ecosystem.

The Challenge for Professionals

In response to the widespread failure to achieve human rights, people's demands have become more insistent. This overriding, global demand for greater production and wider distribution of the eight basic values, transcending all cultures, is the *human rights movement* (McDougal et al. 1980), and we must ask whether events are leading us closer to the goal of sustainable human dignity and a healthy environment or further away. Realistic knowledge of trends and conditions permits us to locate ourselves in this situation and make some decisions about what to do next. Knowledge of the past, good judgment, and creative thinking can help us produce developmental constructs of probable futures.

As discussed in Chapter 5, Lasswell and McDougal (1992) posited a construct of a "universal public order of human dignity," a society that is genuinely democratic, with the rights of individuals fully protected and open to wide expression. Table 8.1 compares this scenario with its opposite, the "garrison-prison state," with regard to all eight value classes. The changes in perspectives and in society in the past two hundred years largely resulted from the explosion of enlightenment; this construct assumes that knowledge will continue to be accumulated and widely shared. Throughout the twentieth century guarantees of human rights were extended to more and more people through the establishment of labor and child-labor laws, women's suffrage, the civil rights movement, antidiscrimination laws, and many more efforts. In this construct, elites will come to recognize that social process is inclusive and open to all. The current trend toward a more universal conception of social process is leading to international law (that is, a system of sanctioning) and institutions to administer it. A common identity in the world community will bring a rational force to public order. As I mentioned, evidence exists for this projection, but its eventuality is not assured. Brunner (1994), for example, suggests that accumulating tensions in the American political system bring its sustainability into question and that we need to restructure it to reinvigorate public trust, a task that both professionals and political leaders could help bring about.

Table 8.1. Comparison of How Values Are Shaped and Shared by People in Two Systems of Public Order

Value	Commonwealth of human dignity	Garrison-prison state
Power	Recognition of the individual as a human being; admission to national membership with full participation in processes and effective power of government; freedom to establish and join groups (political parties, pressure groups, and private associations); freedom of movement across territories; freedom to exercise various strategies; protection of equality under law	Denial of nationality and state protection, causing stateless, transient persons (including refugees); exclusion from office holding, voting, and democratic participation; suspension or manipulation of elections; one-party rule; restrictions on movement and access; suppression of minorities, opposition parties, or political nonconformists; arbitrary arrest, imprisonment, and torture; restrictions on freedom of association and assembly; closed government institutions and processes; mass expulsions of resident aliens; slavery, apartheid, and caste; unwarranted states of siege, national security, or martial law
Wealth	Maintenance of high levels of productivity; enjoyment of benefits from the wealth process based on contribution; freedom to accumulate and use resources for productive purposes; freedom from wasteful uses of resources; participation in groups and institutions in the wealth process	Widespread poverty; lack of basic income and social security; mass unemployment; disproportionate benefits in relation to contributions; forced labor; arbitrary deprivations of wealth, especially during crises; limited job mobility; no freedom of association as in unions; no private ownership; disparities in the distribution of wealth; inflation; overconcentration of wealth; wasteful use of natural resources, depletion without regard for future generations, or diversion for destructive purposes
Enlightenment	Basic education for all plus additional access to further education based on capability and contribution; freedom in the use of language, in local and mass communications, and in assembling resources for enlightenment; freedom from discrimination in the acquisition, use, and communication of information;	Illiteracy; limited opportunities for education; indoctrination through thought control, brainwashing, and conditioning; fabrication and dispersal of misinformation; withholding or suppressing information necessary for people to appraise government policies and decisions; politicization of universities; manipulation of the media and

Table 8.1. continued

Value	Commonwealth of human dignity	Garrison-prison state
	freedom from conditioning or brainwashing or from distorted communications	mass communication; restrictions on freedom of opinion and expression; suppression of languages; keeping people ignorant by restricting travel and communication; suppression of dissenters and nonconformists through coercion; manipulation of tatse and style; censorship
Skill	Acquisition and exercise of basic, socially acceptable skills for participation in all value processes; opportunity for additional acquisition on talent and motivation; opportunity to have talents discovered and to participate in skill-related groups and institutions; special assistance to overcome handicaps; opportunity for training appropriate to a culture of science and technology	Requisitioning of talent and skill; lack of opportunities to develop skills; loss of skill and job choices; compulsory job assignments; alienation from work through automation; restrictions on freedoms of skill groups to organize and operate; obsolescence of skills from rapidly changing technologies; "brain drains"
Affection	Basic acceptance by one's community and capability for effective functioning in it; good family and friendly community relationships; freedom to give and receive affection and to share intimate personal relationships; freedom of association and loyalty toward groups of one's choice	Enforced loyalty to the state; regimentation enforced by power institutions; destruction of rival groups; calculated use of hatred; restrictions on freedom of association; family crises during rapid change; involuntary marriages and adoptions and prohibition of interracial or interreligious marriages; restrictions on terminating bad personal relationships; social ostracism; stifling of personal relationships, especially unconventional ones, by informers or secret police
Well-being	Basic minimum in health (psychological and physical), right to life, safety, comfort; additional opportunities based on choice; development throughout life; merciful euthanasia (freedom to depart or continue	Disease, hunger, starvation, high mortality rates, low life expectancy; death by violence; death penalty; inadequate safety, health, and comfort; stunted development from malnutrition; high levels of mental and emotional prob-

(continued)

Table 8.1. continued

Value	Commonwealth of human dignity	Garrison-prison state
	life); environments that support survival and development; group survival and development; opportunity to benefit from pertinent scientific and technological advances; use of appropriate strategies to prevent disease or injury and restore health; freedom from forced experimentation; access to medical services, transplants, birth control, and genetic engineering	lems; intense anxieties from threats of violence; mass killings; high conflict; genocide; torture; lack of or unequal access to medical services and care, especially for handicapped and aged; poor and crowded living arrangements; degraded environments; inability to cope with natural disasters; lack of family planning, birth control, and euthanasia
Respect	Fundamental freedom of individual choice and mutual deference with regard to choices about participation in all value processes; effective equality of opportunity, with no discrimination based on race, color, sex, religion, political opinion, language, or other grounds; recognition of contributions; a social environment that protects the widest possible freedom of choice and minimizes any kind of coercion; freedom to establish and change identifications; freedom from forced labor, imprisonment for debt, and all acts of violence	Absence of individual freedom of choice, including social discrimination, caste systems, slavery, and equivalent institutions; discrimination; lack of individual autonomy, especially privacy; terrorism; forced labor and debt bondage
Rectitude	Full participation in responsible, morally upright, principled conduct in all social processes; maintenance of civic and public order in which all people demand of themselves and others that they act responsibly for the common interest; freedom to choose secular or religious grounds for responsible conduct; freedom to initiate and join groups and institutions specialized to rectitude	Politicization of ethical standards; enforced conformity in behavioral standards; denial of religious freedom; atheism as national policy; persecution of heretics, nonbelievers, and minorities; warfare over religious differences; compulsory conversion, education, or worship; arbitrary restrictions on kinds or places of worship; rejection of conscientious objection to military service

Source: McDougal et al. 1980.

The obstacles that would limit attainment of human dignity must be addressed, and there is only one avenue open to us: communities everywhere must set up and maintain comprehensive, authoritative, constitutive decision processes to address human rights, population, and natural resource problems. Decision process is the means by which people can achieve freedom, security, and public order, including sustainable management of natural resources.

Such a constitutive process should reflect human rights and should be effective in securing them. All people who are affected by, or who can affect, decision process should participate, and all values should be represented equitably. Every individual should be accepted as important. Responsibility should be central, and all participants should be held accountable for their behavior. Perspectives should emphasize interests that affect all people, rejecting special interests. Demands should ideally include all the values that bring dignity to people. Identifications should be with the world community and all its necessary supports (including a healthy environment and other life forms). Expectations should be realistic and aimed at improving inquiry and communication.

The arenas or organized situations within which participants operate should be set up and maintained for maximum access by all individuals and groups. They should not be unduly bureaucratized but should balance centralized and decentralized concerns. Arenas should be continuous rather than sporadic and their members should anticipate crises. Participants should use bases of power that are pluralistic in terms of authority (based on the shared expectations of community members) and effectively arranged. People should choose strategies that integrate diplomatic, ideological, economic, and military instruments (with a preference for persuasion and against coercion, the last resort for securing public order). Strategies should be designed in detail to promote human dignity with minimal violence, invasion of privacy, and so on, with an emphasis on enlightenment.

Michael Reisman (1969) and others (for example, Lasswell 1971c, e) have described sustainability from a functional, policy sciences standpoint as increasing participation in the shaping and sharing of all values in and among communities of the world. Noting that all countries are involved in some process of international development, Reisman (1969, 2) points out that the optimal community goal is not development in the sense of "achieving a static capitalization and allocation of values conforming to transient demands, but rather the establishment of a viable ongoing development process which is both responsive to environmental changes and challenges, capable of reformulating goals to meet them and able to perform all the other decision functions indispensable to the maintenance of satisfactory community order."

All development takes place as part of a social process that can be inventoried in terms of the eight value categories (Reisman 1969; McDougal 1992–93). Many conventional studies of development and interventions in communities focus only on a few values, however, power and wealth being the two most common (under the assumption that if people had more power or money, all would be well). But the policy sciences demand a focus on the interrelationships of all the values simultaneously since development or social change can influence each positively or negatively. Sustainable development supports the eight values only under certain conditions (see table 8.1). Looking at sustainable development functionally highlights the fact that it will become increasingly important to consider problems and solutions in terms of the global population, global as well as local demands for value-shaping and -sharing, and the global pool of resources. It will be in the common interest of all people to establish and maintain a worldwide constitutive decision process through which environmental problems might be solved.

Sustainable development, in turn, depends on what Leonard Doob (1995) calls *sustainers,* individuals who support and manifest sustainability. People whose personality, motivations, attitudes, knowledge, and actions support this concept or potentially do so must be discovered. For example, renunciation is indispensable; clearly, individuals must forgo immediate gratification for the sake of the future. Doob also looks for adequate knowledge about causes, interdependence, and remedies for sustainability problems, sensitive attitudes toward nature, life, and people, controllability of individual and group actions, and integration of sustaining patterns. Similarly, Lasswell (1951a) describes a "democratic character," a person who has a sense of responsibility and high standards of right conduct applied to public matters, who helps other people gain access to the values on which individual merit rests, and who tries to maintain optimum physical and mental activity throughout life. Both Doob and Lasswell note that society could greatly improve its educational systems to raise more such citizens (see also Habermas 1987; Dryzek 1990).

Achieving sustainability will require a shift in myth, which, as discussed in Chapter 2, is the most fundamental assumptions accepted by a community (mostly as a matter of faith) and a major organizing principle of all human communities (for example, Lasswell 1950a, b; Rappaport 1979; Brunner 1994; Michael 1995). Doob (1995, 5), like many others, notes that the current myth dominant in Western cultures contains many nonsustaining ideas. Among them are the notions of "us against the environment," "us against other men," "it's the individual that matters," "we have and should strive for even more control over the environment," "we live within an infinitely expanding frontier," "economic determinism is common sense," and "technology will do it for us." Slotkin (1992), Michael (1995), and others have also contributed to this list of unsustainable mythic elements, adding beliefs such as "progress," "human domination," and "economic growth." Modern, unsustainable use of the environment has been blamed on the monotheistic, Judeo-Christian tradition, a philosophy that led to Cartesian dualism (people separate from nature), positivism in science, neoclassical economics, and more (White 1967; Turner 1980). Clearly, elements of the dominant myth must change if sustainability is to become reality (see Olsen et al. 1992). Ongoing social processes and policy debates are relatively peaceful means by which to achieve this, but change must come in time and with minimal disruption to the human enterprise.

In order to change, people everywhere will need a much better sense of what their common interests are and how to achieve them. McDougal and colleagues (1980, 409) observed that "one of the most obvious facts about the interactions in contemporary global social process is that all these interactions, whatever their importance or unimportance and however differentially they may affect different individuals or groups, are entirely collective or inclusive in their impacts upon subsequent interactions." Local processes and actions are necessary but not sufficient for achieving global goals; global actions are needed as well. Dryzek (1990) points out that as human populations grow, as natural resources are depleted, and as people interact in increasingly complex ways, the common interest resources (for example, atmosphere, soil, oceans, biodiversity) become ever more important to private special interests. "The continuing integrity of the ecological systems on which human life depends could perhaps be a generalizable interest par excellence" (55). All participants in the global social process, as well as regional and local ones, are likely to have both generalizable (common) and particular (special) interests in any given issue, but as Lasswell and McDougal (1992, 147) observe, "What any one individual can get for himself, and for those with whom he identifies, is in the long run a function of what all other individuals can get." Humankind is characterized by a pervasive interdependence, and there is only one acceptable, genuine way to clarify common interests—by authoritative, democratic decision making. The alternative, a resort to naked power, is not acceptable.

Protecting the environment is within the reach of our creativity. Although some might disagree, environmental destruction is probably not yet at catastrophic levels globally. The perspectives, institutions, practices, and technologies that got us into this crisis might yet be reconfigured toward getting us out of it (see Ham and Hill 1986). We will need to carry out many planning and development projects that will serve the goals of maintaining a secure environmental base and promoting and augmenting human dignity values, which will, in turn, require many delicate and continuing adjustments at all levels of government (McDougal 1992–93). For policy-oriented natural resource professionals, the task ahead is to contribute constructively to decision processes and to turn around the many existing processes and practices that are not sufficient to secure these basic goals. We have enough general knowledge about the scope of weaknesses in decision processes and the conditions and causes of global environmental crises to be somewhat encouraged that solutions are possible. A major target should be shaping the perspectives of effective elites in key institutions so that more rational and efficient constitutive decision processes can be set up to secure our basic goals.

Learning from Policy Experience

We know what the goal of policy should be—to create and maintain systems of public order that embody human dignity—and we know that for this to happen we need to ensure a sustainable flow of natural resources. The best way to work toward this goal is to develop skills in the policy sciences' problem-solving approach and to learn better from experience. The key is to find the lessons of hindsight, translate them into guides for future policy, and overcome resistance to genuinely interdisciplinary approaches and policy learning.

Learning is the process of using information to adjust our responses to the environment; it is the process of detecting and correcting "errors," that is, mismatches between practices and outcomes (Argyris and Schön 1978; Davis et al. 1988; Rose 1993). Learning to meet both human dignity goals and natural resource goals involves much more than just refining or redoubling our commitment to positivistic scientific or conventional methods. Instead, we must focus on our own learning capabilities. The challenge is *to learn how to learn* most effectively (Etheredge 1985). That is, we need an approach that improves our policy performance by explicitly seeking information about our own past performance, the dynamic status of the problems we face, and the contexts of these problems. This means we must look at ourselves and our organizations. Thus, *policy learning* can be defined as a process in which individuals apply new information and ideas to policy decisions (Hall 1993).

The policy sciences framework offers a vehicle for systematic policy learning—a shared, stable frame of reference within which to appraise policy experience honestly and empirically and to compare policy experiences and processes functionally in order to find patterns and lessons. In addition, it offers a language with which to communicate lessons efficiently and effectively. We can use the framework's categories and their relationships in any particular case to gain insight. For example, how does a change in participants or perspectives affect how values are shaped and shared in the social process? We can compare how each of the decision functions is carried out, and with what effect, across a number of cases to gain understanding about large patterns in policy behavior. The framework forces us to think critically and responsibly about natural resource conservation in support of human dignity. If we learn well and quickly, then numerous possibilities for a better world open up.

The challenge in learning is not just to break down or "deconstruct" policy process or content but to see connections between the two and understand the networks of interactions from which policy results. Policy-oriented learning, which may require adjusting the belief system of individuals and changing the structures, procedures, and customs of organizations, can lead to significant alterations in thought and behavior (Sabatier and Jenkins-Smith 1999). An explicit learning strategy demands systematic inquiry, changes in direction, new ideas, bridging, and shared responsibility for outcomes.

Individual Learning

Learning can happen at three levels—individual, organizational, and societal—and learning at one level may affect learning at the others. Individual learning is essential to personal and professional change as well as to organizational and societal learning (see Parson and Clark 1995); policy learning is only possible if individuals learn first. Several authors argue that the common, natural process of reflection is the best way to learn (for example, Schön 1983). We reflect when we analyze and discuss problems. This process, often stimulated by a nagging, unresolved problem or challenge, involves asking questions. Whether it occurs more readily as a solitary or a group activity is unknown, but it takes place in both settings. Often we experience breakthroughs in our thinking, or insights, while involved in some other activity. One barrier to reflection is that people place a higher value on action; the expectation for professionals in conventional settings is to act, not to think or reflect. Probably the single most important element in learning to reflect more effectively—and that which many professionals lack—is having time to reflect. We all can learn from experience by systematic reflection if we take the time to do so. It is an inexpensive way to learn that can be studied and fostered. Professionals can counter the fast pace of change confronting them by developing a systematic capacity to learn from current work settings and adapt lessons to new contexts. The categories in the policy sciences framework are particularly helpful in extending the breadth and depth of our reflection.

A number of authors have offered advice for stimulating individual learning, but it is applicable at all three levels. Donald Michael (1995) stressed building and employing a more useful vocabulary of metaphors and problem-solving concepts to reflect reality more accurately. We often use sports and war metaphors that subtly emphasize "we-they" parochialism and "winner-loser" divisions that poorly map complex human societies and dynamic ecological environments. The use of such loaded metaphors is widespread among activists, professionals, decision makers, and the general public when thinking about problems and policy situations. But the metaphoric power of language is underappreciated; new metaphors could come from ecology, music, storytelling, and learning itself. We live, as Michael says, in "an amorphous, problematic, information-rich world of multiple myths described by such words as reciprocal, resilient, circular, emergent, development, ebb and flow, cultivate, seed, harvest, potential, fittingness, both" (477). Michael also suggests minimizing learners' sense of vulnerability as they develop their knowledge and skill in practical problem solving. Individual effort is required, but learning groups could expedite the process. Learning is more successful when it takes all the significant policy issues, not just the "facts," into consideration. This means considering all perspectives, including peoples' myths, the dynamics of disciplines and bureaucratic turf defense, and political expediency, among other things.

Gareth Morgan (1986) recommends encouraging and valuing openness and reflectivity as a way to upgrade policy learning. These attributes are essential in a world of uncertainty where life is always complex and environments are ever-changing. Morgan also recommends exploring different viewpoints and avoiding imposing structures of thought and action on problematic situations. Ron Westrum (1986) suggests a number of principles for improving creative thinking that are applicable to larger policy arenas: Encourage systemwide awareness in all members of the policy system and creative and critical thought in all organization members, he says, and link the parts of the system whose work is interdependent. In addition, reward communications and activities that show a desire to contribute to the entire system's thought processes, avoid overstructuring problem solving, and examine mistakes honestly.

Organizational and Societal Learning

Changing how organizations learn is also paramount. Organizational learning is based on individual learning, and societal learning is focused on how people learn in organizational contexts (Argyris 1992). Often individuals know far more than is institutionalized in the organizations for which they work. That an individual learns something does not mean that co-workers or the organization itself benefits. Organizational structures, procedures, and customs either permit or restrict both individual and organizational learning (Popper and Lipshitz 1998). Organizational practices may not change even when lessons for improved performance are brought to the attention of leaders (Clark 1996b, 1997a). Organizational learning is more difficult to bring about than individual learning, but if learning processes can be institutionalized, they can be carried forward to individuals and even across generations. Societal learning, the highest level, occurs as individuals and then organizations improve their policy responses. The capacity for societal learning is a direct function of the kinds of organizations a society creates and maintains. Open, flexible ones that promote learning, adaptation, and change are ideally suited for policy learning at the societal level, whereas closed, rigid ones resist learning.

Another way to learn is by paying attention to "focusing events," or crises that capture policy makers' and the public's attention. These incidents can drive the process of learning and change at all levels (Birkland 1997). For example, the 1989 spill of eleven million gallons of oil from the *Exxon Valdez* in Alaska riveted people's attention, clarified people's expectations about what was acceptable, set up an opportunity for discussion and learning, and eventually led to changes in energy production and transportation systems (see Perrow 1984). Such focusing events, regardless of the issue, can be analyzed using the policy sciences framework to learn comprehensively, productively, and constructively (see Reisman and Willard 1988). Without a framework for learning it is possible for people to learn wrong or incomplete lessons, to focus on technological solutions alone, for instance, without also seeking improvements in social process, decision making, and problem orientation.

It is clear that policy learning and change come about when networks of individuals and organizations in a given policy domain are structured and function in ways that enable or facilitate learning. Learning is usually a long-term effort, although focusing events may accelerate the process. It usually occurs incrementally, but, like evolution's "punctuated equilibrium," it may also occur dramatically when an event focuses attention on the need to change. Learning may also take place through comparison and reflection when one policy domain is exposed to problems and solutions in other arenas. It is strongly shaped by organizational context; it can be hastened by procedures that produce information and lessons that are acceptable to potential policy learners (Senge 1990). Policy learning can also take place though appraisal of decision processes. The categories of the policy sciences framework can be viewed as variables to be researched and manipulated, thus creating a system of continual learning and adaptation in response to experience over time. This is the true meaning of adaptive management.

Policy learning at all three levels—individual, organizational, and societal—should seek to improve the substance, process (or governance), and structures entailed in policy (Brunner, cited in Clark 1999). Substantive issues might include more sustainable ways to use natural resources (for example, forests, grasslands, and oceans). The governance aspect of policy learning can focus on ways to improve patterns or processes of participation, open debate, data acquisition, planning, implementation, and evaluation for policy purposes. We can also improve the structures of policy by learning about interpersonal and organizational structures for research, practice, and education. Although some policy learning occurs at all three levels in many arenas, accelerated, systematic learning is desperately needed to secure advances in natural resources and other contexts.

These approaches to learning how to learn all call for a process of open-ended inquiry, that is, remaining open to environmental changes and challenging operating assumptions in a basic but constructive way. Such an understanding is liberating and can greatly enhance your problem solving.

Barriers to Policy Learning

There are a number of barriers to policy learning and problem solving, some of which I mentioned earlier (see Miller 1999). First are certain attitudes and characteristics of individuals (Putt and Springer 1989; Mares 1991). Closed minds do not learn. Some individuals are predisposed to new ways to solve problems and are ready to learn about the policy sciences. Others resist change, lack interest or commitment, or orient themselves toward divergent views, some rooted in rigid ideology (Barber 1961). Some people fear a loss of status, prestige, power, job security, self-esteem, or simply the unknown; forced changes threaten their work philosophy and practice. These psychological constraints are hard to overcome. Part of this barrier is disciplinary boundedness. Professionals are usually trained in and loyal to a particular field, and such outlooks are deeply ingrained in university education and on-the-job requirements. Disciplinary education is a major means by which professionals bound their rationality, since most disciplines do not teach basic conceptions of learning or broader identifications, and they do not train practitioners to be open to alternative problem-solving means (Schön 1983).

Putt and Springer note that difficulties in communicating policy and program goals, progress, and results constitute a second class of barriers. If people do not have a common frame of reference in policy work, then it is difficult to communicate effectively, and if there are poor collaborative relationships, then policy information is likely to be devalued. The policy sciences were designed to improve communication, as well as policy and programs, by focusing on the diverse information needs of interdisciplinary research. The technical quality of policy work also affects its subsequent use, so it is important to use the best conceptual tools. The policy sciences offer a vehicle to ensure high-quality analytic work and communication and to avoid advancing policy options that are philosophically, technically, or practically weak.

Third, certain characteristics of the policy "landscape" also constrain policy learning. What is learned, when, and how is determined largely by contextual and situational factors. The structure and design of organizations, for instance, affect how people understand and solve policy problems since organizations have built-in policy preferences and modes of operation that precondition employees to see the world in certain ways and conduct their work according to certain operating procedures. Because of this, some organizations are better at policy learning than others. Political climates and economic conditions are also important; it is easier when organizational, political, and economic factors all support policy learning. Manwell and Baker (1979) note that the "establishment" may resist alternative ideas for improving policy learning for a host of reasons, including personal conflict, informal and formal information flows that prevent messages from getting through, professional territoriality, low tolerance for diversity and dissidence, a fickle reward system, in- and out-group differences, closed-mindedness, and a tendency to rationalize or ignore injustices that eventuate. Together, these three sets of factors make up a formidable array of cultural and social sources of potential resistance to policy learning. They must be understood and surmounted if policy learning is to be improved at the individual, organizational, and societal levels.

Conclusions

Throughout this book I have discussed the role of professionals in solving natural resource problems. In the face of accelerating environmental losses and breakdowns, gridlocked political and bureaucratic machinations, disciplinary specializations, organizational changes, and conflicting demands, it is increasingly difficult for professionals to know what their role is and how to perform it well. Clearly, the accumulation of similarly unsatisfactory results and outcomes is evidence that our usual problem-solving methods do not work. But, as Jack Ward Thomas (1986) has written, the hallmark of professionalism is effectiveness, and to be effective requires us to use the very best means available to solve society's problems. There is too much at stake to do less.

The methods introduced in this book offer you a way to integrate information—from many sources, in diverse forms, about content and context—in order to develop sound judgment in your problem-solving efforts. These methods also help you to comprehend political situations, that is, the competing demands, claims, and maneuverings for advantage, and to develop responsible influence in them. In addition, they are a constant reminder of the ultimate goals of our policy-making and management efforts—to clarify and secure common interests and to manage natural resources in ways that serve the public good, that is, that promote dignity and well-being for all people. Taken as a whole, they constitute a new standard for critical thinking, problem solving, and professionalism (Lasswell 1970). It is my hope that this book has served as a useful introduction to these conceptual and practical tools and as an invitation to all natural resource professionals to learn and apply them.

SUGGESTIONS FOR FURTHER READING

Brunner, R. D. 1996. A milestone in the policy sciences. *Policy Sciences* 29:45–68.

Brunner, R. D., and W. Ascher. 1992. Science and social responsibility. *Policy Sciences* 25:295–331.

Chen, L. 1989. *An introduction to contemporary international law: A policy-oriented perspective.* New Haven: Yale University Press.

Clark, T. W., A. R. Willard, and C. M. Cromley, eds. 2000. *Foundations of natural resources policy and management.* New Haven: Yale University Press.

McDougal, M. S., H. D. Lasswell, and L. Chen. 1980. *Human rights and world public order: The basic policies of an international law for human dignity.* New Haven: Yale University Press.

Sachs, A. 1995. Eco-justice: Linking human rights and the environment. Worldwatch Paper 127, Washington, D.C: Worldwatch Institute.

Appendix: Interdisciplinary, Problem-Solving Workshops for Natural Resource Professionals

Tim W. Clark, Robert J. Begg, and Kim W. Lowe

This appendix describes a multi-year effort to teach the policy sciences problem-solving approach to experienced natural resource professionals in Australia. These interdisciplinary workshops were carried out to meet a growing demand for more effective professional performance.

Calls for better problem solving in natural resource conservation and management are being made in many different arenas (for example, Heberlein 1988; Cole 1992; Shrader-Frechette and McCoy 1994; Christensen et al. 1996; Clark 1996b; Lubchenco 1998), motivated in part by the increased complexity of today's challenges compared to those of just a decade or two ago (see McDougal et al. 1988; Kates et al. 1990). Haeuber and Franklin (1996, 692) note that "few areas of public policy are as contentious as the issues surrounding management of our environment and natural resources." Professionals who use problem-solving methods and "cognitive styles" that are outdated or less than effective invite failure (Miller 1985; Johnson 1992), and faltering programs cannot be tolerated because they result in species extinctions, endangered ecosystems, loss of topsoil, and air and water pollution, to mention only a few problems. Former United States Secretary of the Interior Bruce Babbitt, for example, called for development of an "interdisciplinary science" to meet today's natural resource challenges. But changes will be needed in education in order for practitioners as well as university students to become interdisciplinary scientists. These changes, "as obvious as they are essential," require a move away from conventional, discipline-based education toward interdisciplinary, problem-solving knowledge and skills (Hein 1995, 84; see also

Lasswell 1971a; Clark 1986a, 1992; Heberlein 1988; Foundation for Strategic Environmental Research and The Council for Planning and Co-ordination of Research 1998).

One way to develop such skills with potentially long-lasting benefits is through workshops (Brewer 1973c; Hanna 1994). Such exercises can help professionals move beyond narrow, technical outlooks by articulating and strengthening new ways of synthesizing, thinking, and solving problems (Sullivan 1995). In this appendix we describe three workshops and two years' follow-up designed to teach interdisciplinary problem-solving methods. We also evaluate these workshops as a vehicle for learning and applying these methods both inside and outside government.

The Workshops

Workshops were conducted for staff of the Flora and Fauna Branch (FFB) of the Department of Conservation and Environment (DCE) in Victoria, Australia, for which Robert Begg was manager of the wildlife section and Kim Lowe manager of the threatened wildlife program. The workshops, led by Tim Clark, consisted of lectures, readings, case analyses, presentations, and discussions, and were based on his courses at the Yale School of Forestry and Environmental Studies since 1990. They also included exercises on how to apply this new approach to selected wildlife, flora, forestry, and ecosystem management challenges that this agency faces.

Participants and Schedule

The workshops were held at various DCE offices in and near Melbourne. Nineteen people participated in the 1994 workshop on 3–4 and 8–9 November, fifteen in 1995 on 29 November–1 December, and twelve in 1996 on 28 November. These individuals, averaging about fifteen years' experience (range seven to thirty-five years), included a branch manager, mid-level managers, research scientists, planners, policy analysts, heads of the endangered species and ecosystem management units, and a university professor. Six of these participants were contacted in 1997 and 1998 and interviewed concerning long-term effects of the workshops.

In 1994 the first two days were devoted to lectures and discussion of contemporary professional practice and conservation challenges that professionals face. We identified and discussed problem-solving concepts and methods that participants were currently using. These were described in "conventional" terms using everyday language and had been learned partly from university training but mostly from on-the-job experience. In fact, many people did not recall explicitly learning any kind of generalized problem-solving method or even applying one in daily practice. Participants recounted numerous actual cases in which they had been involved, ranging across many wildlife and resource issues. Referring to actual experiences helped the participants to clarify, articulate, and understand generalized conceptual models that they had used. Describing these methods, comparing them, and appraising their success resulted in extensive discussion.

Participants varied in their abilities to reflect on their own problem-solving approaches and to describe them with specificity and sophistication. It was difficult for some individuals to raise to conscious attention the methods they used and to conceive just how they "solved" problems. We discussed these individual thought processes as well as concepts of how the policy-management process works in organizational contexts and how we would like it to work—subjects that had never before been addressed systematically by most participants. Many found that they actually shared certain perceptions and problems but did not realize it before these discussions.

Next, the policy sciences framework used throughout this book was introduced (see fig. app. 1). This framework, with its brief set of categories, systematically guides attention to key variables in problematic situations. This method was contrasted with the approaches participants had been using, and we found that several key variables—especially social process or context—had been underappreciated or overlooked when conventional approaches were used.

In the second two-day session, we applied the new method to a wide range of actual management cases, programs, and professional issues. Teams of two to three people analyzed and presented each case using the new method. This exercise gave people a greater understanding of the method and experience in applying it. Each presentation stimulated further discussion, focusing on learning from past experience and improving future management. This hands-on exercise proved highly instructive. Some participants seemed trapped by convention, unable to step outside their old concepts and methods, and could not incorporate the new concepts, vocabulary, or functional thinking into their outlooks. Some simply could not think reflectively or abstractly about their years of work. Others quickly grasped the new method and applied it with facility. At the end of the workshop, participants evaluated the exercise in writing.

The 1995 workshop was held over the course of three days. Fifteen

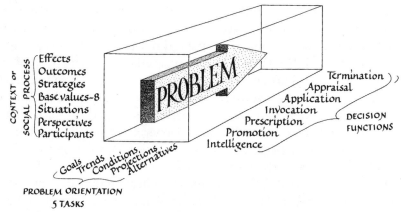

Appendix Fig. 1.   A simplified illustration of the natural resource management and policy process. The three axes show the three key dimensions of interdisciplinary problem solving—problem orientation, context or social process mapping, and decision process analysis.

people participated, six of whom had been in the 1994 workshop. Again, considerable work experience and a diversity of jobs were represented. A member of the nongovernmental conservation community also attended.

The 1995 workshop focused on the interdisciplinary problem-solving method itself, with less emphasis on applications. If participants were well grounded in the method, it was thought that they could apply it themselves to whatever management problem they faced. We again systematically examined the framework and its integrated concepts and categories. As in the 1994 workshop, the method was illustrated with case examples from the United States and Australia. These exercises reinforced what had been learned in the lectures. There was much discussion of the method and its application, and again the group completed the workshop with a verbal and written appraisal.

In 1996 the day-long workshop of twelve people was similar to the previous two. The method was briefly reviewed and then general discussion ensued. Again, participants were asked to evaluate the workshop and the method.

Readings and Cases

The 1994 and 1995 workshop participants were given readings to study in advance and were asked to be prepared to discuss their standpoint, to share stories about their work experience, and to familiarize themselves with at least one case history to bring to the discussion.

The readings were integral to both workshops. The 1994 readings included twenty-four articles totaling about 400 pages. Four readings were devoted to each of the decision functions. Some papers addressed key concepts and others covered practical applications. The instructor summarized each paper, and the group discussed the main points as time permitted.

The 1995 readings included forty-four papers totaling about 400 pages. Except for description of the method itself, there was little overlap in the two sets of readings. The 1995 readings were divided into six parts. Part I addressed the demands of professional practice. Part II was an overview of the interdisciplinary problem-solving method, and part III was a more detailed treatment. Part IV looked at the issue of "defining" problems, part V examined other tools for interdisciplinary problem solving, such as prototyping, chart rooms, and decision seminars, and part VI was a conclusion and a look to the future. Because time was limited, the readings were summarized and discussed.

The readings were designed not only as background material, but also as reference sources once the workshops were over. Not all the readings were self-explanatory or meaningful by themselves, however; discussion was needed to appreciate their relevance. Some participants returned to the readings after the workshop and read all the papers in detail, and others drew on a few key papers in the course of their daily work. The papers on problem orientation and problem definition were especially well used.

The participants best discovered the power and utility of interdisciplinary problem solving by applying it to actual situations. Learning was especially quick for those who had spent years on a particular case. They became skilled in the use of the new method by moving between theory and application. Participants learned from one another as they listened to the presentations, listened to others analyze cases, and discussed issues. This substantive interplay, discussion, and joint problem solving illustrated the policy sciences approach and enhanced communication and other skills necessary for effective conservation.

All the participants in these workshops had considerable experience in their professional careers. Some had spent many years on some issues and were able to offer case materials as examples and as objects of detailed analysis. Each small team was asked to demonstrate a command of the method by applying it to a real case, or in other words, to reframe their case studies within the terms of the new method. This was the principal means to determine if they understood the concepts and categories. They were asked to be both descriptive and prescriptive—that is, to describe the problem and its associated social process systematically and to recommend what was needed to resolve it. During and following each presentation, the other participants provided critique and discussion.

Key lessons from each presentation and discussion were listed on the chalk board and compared at the end. Among the lessons derived from the case analyses were (1) that management "problems" are usually much more complex than managers initially think, (2) that managers need to remind themselves constantly of context, that is, the social process surrounding issues, and they need to use this information better, (3) that more attention should be directed at constructing good problem definitions, and (4) that all parties involved need to make special efforts to communicate better.

The 1994 cases included management policies for urban possums, orange-bellied parrots, Leadbeater's possums, brolgas, koalas, kangaroo culling, emu farming, the issue of lead versus steel shot in hunting, and the department's Land for Wildlife program. The 1995 cases (many of which were reviewed in the 1996 workshop) included spotted tree frogs, black-eared miners, New Holland mice, regent honeyeaters, cockatoos, live fish reintroductions, tree fern and sphagnum moss harvesting, nature conservation on private lands, the Land for Wildlife program, establishment of a Central Highlands sanctuary, management of remnant grasslands, human safety regulations, uses of the wildlife atlas database, and reservation of different habitat types. Koala management policy in Victoria was examined as a group exercise, and existing and proposed management plans were critiqued. Suggestions for improvement in this case included ways to clarify the plans' goals and appraise existing science and trends. The group also came up with several management alternatives that were judged technically and politically feasible but had not been included in the original plan. These were contextually tailored and detailed for each case.

The Interdisciplinary Problem-Solving Method

The workshops were designed to help professionals solve problems practically and think more efficiently, effectively, and responsibly about public policy. Our workshop taught an explicit, genuinely interdisciplinary method with an integrated set of concepts and categories for thinking critically about problems (see Lasswell 1970). These methods, described more completely in this book, are summarized briefly here. Most other educational approaches to broadening the scope of biological and natural resource professionals merely expose people to a diversity of disciplines, such as sociology and political science, and leave it to the individual to integrate them, if they can. The policy sciences are explicitly integrative (Lasswell and Kaplan 1950; Lasswell 1971a; Lasswell and McDougal 1992).

Founder Harold Lasswell felt that interdisciplinary problem-solving requires problem solvers to do three things, all the while remaining aware of their own standpoint in the problem-solving process (table 1). The first is problem orientation. People must orient themselves to

Appendix Table 1. Principal Conceptual Tools in the Interdisciplinary Problem-Solving Method of the Policy Sciences

Maximization Postulate: All living forms tend to complete acts in ways that are perceived to leave them better off than if they had completed them differently.	
Social Process	Components
Participants	Individuals, groups, value shapers (official, non-official), value sharers (official, nonofficial)
Perspectives	Identities, myths (doctrine, formulas, mirandas), expectation, demands, value demands
Situations	Unorganized (territorial, pluralistic), organized (territorial, pluralistic), value inclusive or exclusive, crisis or intercrisis, in terms of space, time
Base Values	Positive assets (perspective, capabilities), negative assets (perspectives, capabilities) by the eight value categories (see below)
Strategies	Coercive, persuasive, communicative (diplomacy, propaganda), collaborative (military, economic)
Outcomes	Value indulgences, deprivations, decisions, choices (by phases of decision process)
Effects	Value (accumulation, enjoyment, distribution), institutions (structures, function, innovation, diffusion, restriction)
Values	Outcomes and Institutions
Power	Victory or defeat in fights or elections (government, law, political parties)
Enlightenment	Scientific discovery, news (languages, mass media, scientific establishments)
Wealth	Income, ownership transfer (farms, factories, banks)
Well-being	Medical care, protection (hospitals, recreational facilities)

(continued)

Appendix Table 1.   continued

Values	Outcomes and Institutions
Skill	Instruction, demonstration of proficiency (vocational, professional, art schools)
Affection	Expression of intimacy, friendship, loyalty (families, friendship circles)
Respect	Honor, deference (social classes and castes)
Rectitude	Acceptance in religious or ethical associations

Decision Process	Outcomes
Intelligence	Gathering, processing, and disseminating information relevant to decision making
Promotion	Active advocacy of policy alternatives
Prescription	Setting community policy that is both authoritative and controlling
Invocation	Provisional characterization of events in terms of a prescription
Application	Resolution of disputes with regard to a prescription, including sanctions for noncompliance
Appraisal	Evaluation of past decision process, including assigning responsibility
Termination	Ending prescription and arrangements made in accordance with the prescription

Problem Orientation	Questions and Tasks
Goals	What future states are sought in social process? (goal clarification)
Trends	To what extent have past events approximated the preferred goals? (trend, history description)
Conditions	What conditions have influenced the direction and magnitude of the trends described? (analysis of conditions)
Projection	If current policies are continued, what is the probable future of realizing the goals, or what discrepancies exist? (projection of future developments)
Alternatives	What intermediate objectives and strategies will best realize the preferred goals? (invention, evaluation, and selection of alternatives)

Source: Adapted from Lasswell 1971a; Brunner 1995a.

the problem rationally, asking five questions: What are the goals and preferred outcomes? What are the trends in the situation? What factors have caused or influenced this situation? What is the probable course of future decisions and events? And, finally, what alternative options are available to achieve the preferred outcomes, and how should they be evaluated?

The second is social process. The context must be "mapped," or analyzed, so that the practical politics of the situation can be understood. Natural resource problems are always part of a larger human setting, and events make sense only if we understand how they are interrelated. This larger picture of the social process must examine the participants, their perspectives and situations, their values (power, wealth, skill, enlightenment, well-being, affection, rectitude, and respect) and the strategies they use to achieve them, and the outcomes and long-term effects of their decisions and interactions. It recognizes that values are at the center of the natural resource policy process, and thus problem solvers must consider people's perspectives in order to find solutions.

The third is decision process. All decision making performs seven functions: intelligence (planning, or obtaining and processing information), promotion (debate, or recommending and mobilizing support for policy alternatives), prescription (setting the rules, or formulating and enacting policies or guidelines for action), invocation (enforcement, or initial or provisional characterization of how people should behave with regard to a prescription), application (resolving disputes), appraisal (evaluation), and termination (ending a policy or moving on to new ones) (Lasswell 1971a).

Problem solvers must employ a variety of empirical research methods to explore the problem and its context fully. Some methods are extensive, such as surveys of newspapers and general discussions, whereas others are intensive, such as one-on-one interviews and detailed research. Both qualitative and quantitative methods are used. The problem itself indicates what kind of research methods are needed. Different methods yield different results, and people must be careful not to become method-bound. All too often, practitioners become skilled in one set of methods and apply them in all situations, whether appropriate or not. All methods, including observation, questionnaires, qualitative and quantitative assessment, and many others, should be used, according to Lasswell (1971a).

Finally, all this information about problem orientation, social process, and decision process must be integrated. The framework helps professionals see the relationships within this information and guides their judgment in applying it. This method requires some users to enter into a frame of reference that is quite different from ones they might have used in the past. Other users immediately comprehend part or all of this new frame of reference, thus making their transition minimal. Many workshop participants had an implicit understanding of it based on their considerable experience, but not the vocabulary to articulate it. Shifting frames of reference invites a certain kind of intellectual and personality flexibility. Because minds are not blank slates, shifting frames is easy for some people, hard for others, and impossible for still others. Having an ability to abstract from experience is helpful for understanding and applying the new interdisciplinary framework.

Evaluation

Participants were asked to evaluate the workshops in both 1994 and 1995. This enabled us to assess their utility as a vehicle for learning the framework.

1994 Participant Evaluations

Participants were asked a number of questions. First, was the material new and was it helpful to you? Minor corrections have been made in the punctuation of some excerpts.

> "Concepts almost entirely new (although perhaps some had been applied intuitively); certainly the systematic development of the policy sciences framework was entirely new."

> "Perhaps the best 'new' item was a recognition (by me) that problem orientation/identification was the basis for clear process development. Most was not new, but much of it was interpreted differently."

> "For me, it put the process in context and verbalized and mapped out what I innately know. It was interesting in that it showed me aspects of my management problems that I know to exist but either consciously or not, do not address when I solve problems."

Second, what are three take-home lessons for you?

> "(1) The importance of all 5 components of problem orientation—cannot afford to neglect any. (2) The value of regular evaluation of each phase of the policy process, and in particular that the evaluation involve honest critical project appraisal, perhaps by people not directly involved in the project. And, (3) The benefit to a successful outcome of correctly identifying the underlying premises of each of the major players (standpoint)."

"(1) It is definitely worth the time and investment for each of us individually and as a group, to strive for a true 'problem-orientation' to our work. (2) We must pay a lot of attention to problem definition. This must be very sophisticated in its approach. And, (3) Most cases are more complex than they appear initially. We must devote a lot of energy to understanding the context of issues that arise."

"(1) There is a process that can and should be utilized to assist me in problem solving where I am in the 'blizzard' [of details and work]. (2) There are always differing viewpoints, processes, objectives, etc. that impinge on problem solving that need to be considered and included. And, (3) The solution to all problems will be difficult in each case yet many have the same elements as a causal factor. Don't need to reinvent the wheel."

Third, how will you utilize this workshop?

"I think some aspects of the general models may be used selectively and others ignored. I can see one of their main uses in workshops where issues are discussed and it would be very useful and helpful to have them as a model for framing discussions. Personally I think the models are stimulating in encouraging you to think about how you do things, think about and solve problems. Skill in its use will drop out without practice."

"Stick to the two road maps [figures handed out] on the wall when a problem arises—think 'policy process.' Don't feel snowed under when the complexity of the issues I deal with [overwhelms me]—take a bit at a time. Locate project in life cycle."

"I will place the road map for problems in a prominent position in my office. I will select three major problems I am faced with at present and map out each one so I can better understand it. I will assist at least one other member of my staff in mapping out a problem."

1995 Participant Evaluations

Again, participants were asked two questions to evaluate the workshop. First, was the workshop helpful?

"Having little or no prior practical experience with developing policy, the workshop has provided me with a conceptual framework from which to attack any future policy issues/initiation. In addition, it provides me with a tool to assess or critique others' work on policy and in doing so, learn from others' mistakes. I certainly found the workshop helpful and I think that follow-up workshops will strengthen what I've learned."

"Workshop has reinforced the importance of social process in all aspects of decision making process. Methods of problem solving as outlined in the course can actually save time by (a) focusing on the problem, (b) considering all angles, and (c) coming up with practical solutions. If social processes are correctly followed then all participants/stakeholders will have ownership."

"Again it was a most useful workshop. The format of this year's workshop was better than last year—I think we all got more out of you [facilitator] than we did from the 'group' discussions."

Second, what did you get out of it?

"I still have much to come to grips with, the language and basic principles and ideas, however, with future readings and re-evaluation of the principles canvassed over these three days, I'm sure I can 'learn to learn' and use these tools positively."

"While it is good to get this course it is important that it is useful to our work. For this reason, I would like the follow-up of the work-shop to show us where we are doing well and where we made mis-takes so I believe a review day (or days) is essential."

"I will certainly be applying the method to problems and projects: past, present, future to see what it delivers. I feel better about the tools I have to deal with issues in my daily work. Maybe I'll do my job better as a result. [My boss] has already asked for a section-wide briefing. INSTITUTIONAL LEARNING HAS BEGUN!"

Comparison Between Years

Four themes emerged from the evaluations of the workshops. First, analyzing real-life cases helps significantly in generating an under-standing of the interdisciplinary method and in appreciating how each category applies to participants' own work. Second, the workshops left all participants with the clear understanding that science alone is not sufficient to manage natural resources effectively. Third, the method forces professionals to address management problems systematically, explicitly, fully, and contextually. Fourth, all participants agreed that a means must be found to maintain the impetus begun in the work-shops—to continue learning and to improve their skills after they re-turn to the routines of the workplace.

1996, 1997, and 1998 Participant Evaluations

Four questions focused discussion in 1996, 1997, and 1998. First, how and to what extent have you used the interdisciplinary problem-solving method since you attended the previous workshops?

"I have used [this method] consciously at certain stages in two proj-ects in particular—the development of a strategy for managing con-servation reserves, and the design of a reserves database."

"I am writing a book on the theme of corridors, connectivity and conservation. I have consciously used insights from the policy sciences to try to structure the contents and sequence of the book, and also in recognizing the importance of the 'human dimension' in discussing design issues and in making recommendations for conservation."

"I am particularly more aware of the social process and how it impacts the ability to solve the problem. This has benefited me enormously and even more so in my dealing with new clients/ stakeholders in my wider role with the Department."

Second, what can we do to become better problem solvers?

"[Need] wider training in [this method] within [our agency], especially at management level. [Make] the policy framework used in planning individual projects known to (and appreciated by) stakeholders. [And] more conscious application of policy process framework built into organisational processes like work program development and personal assessment."

"Give more attention to clearly defining the problem, and gaining common agreement and commitment to the primary goals to be achieved."

"I found the case of the Koala problem very informative, particularly since many of the problems [we encounter] parallel those in game management where stakeholders have strong and often unchangeable views. I would really like to see this approach taken with some of our projects in a more formal sense."

Third, what issues are you currently working on that could use this approach to problem solving?

"Parks and reserves. General wetlands policy. Forest management. Hog deer management. Urban possum management. Briefing the Minister on diverse issues."

Fourth, what problems have you encountered?

"There have been many distractions and changes since the first workshops. In particular, the reduction in staff has left most people with increased workloads and reduced budgets to manage existing programs. [This has led to] dramatically lowered morale. This impacts on one's ability to complete long-term projects as there are constant interruptions. It is not uncommon for a project which should take five days to complete to actually take up to two months as crises which must be addressed at once erupt."

"After initial successes the Orange-bellied Parrot Recovery Team (Australia's first) was drifting, and the policy concepts have helped generate renewed energy and discipline to the process, refocusing on the goal of recovery in the wild. Similar comments could be made for other Recovery Teams. There is still a lot to be done in these areas."

"I found that the conscious use of [this method] was overtaken by 'too much activity' and seemed to get lost along the way for quite long periods of time."

Recommendations

The workshop participants made recommendations to improve the teaching of the method through in-service workshops as well as its continued use in professional practice. These recommendations generally fell into three categories.

More Repetitions of Case Analyses and Applications

Future workshops should include more repetitions of case analyses. After the workshops are over and participants are back on the job, they should apply the new method to actual cases to practice and enhance their analytic skills. This includes commitments to study the readings, use the handouts (especially the figures) to guide future work, and involve colleagues who did not attend the workshops.

"I would like to see more clearly case histories of the use of this method to flesh out the theory. Examples facilitated discussion and understanding of the application of the method. At a future workshop, I think there could be great benefit in 'working' through some real problems, such as the Koala issue, led by a disciplined facilitator. Again, this would facilitate understanding and expedite adoption of the method in our individual spheres of work."

"All 'policy' staff should sit through this—one of most valuable tools they will learn and have exposure to. We as 'learned practitioners' could benefit from, say, a 'one/year(?)' get-together to present issues we have subsequently dealt with, to review and discuss how well we are implementing/improving policy; what experiences and problems; any updates in policy sciences."

"Read the readings and consider their application to my work. Keep the policy process diagram, and several others in the readings, to hand at my desk to consult when problem assessments are being undertaken. Find/make time to undertake adequate reflection on the problem orientation, social process, and decision process. [Conduct w]orkshops of all of my projects using these processes and involving as many of the players as possible."

More Discussion of Framework's Concepts and Categories

Future workshops and post-workshop activities should include more exposure to the concepts and categories of the framework. Because the method was new to most participants, additional exposure to it and discussion of how to apply it practically would be immensely helpful. It was also recommended that workshop participants discuss the concepts not only among themselves but also with colleagues who did not attend.

> "MORE NEXT YEAR! I can see us needing to review the 'multiple' methods available to see if we can utilize them. . . . Need a pull-out sheet with definitions/and or key words. More examples. Overall, taught professionally, well done."

> "A summary of key concepts and terminology, on a page or two would have helped further (e.g., positivism, myth, etc.)."

> "The [summary given], along with those diagrams should form the basis of the course with a short number of ancillary case examples. The reader as a whole was useful but should not be used as the vehicle for the course—there is too much in it. The workshop has been a great learning exercise for me. I haven't done so much clear thinking since my High School Certificate. The fact that I see societal intercourse in a new light is the best thing for me."

More Agency Support for Improved Problem Solving

Participants generally felt it would be difficult to apply the new method once back on the job. Tradition, agency rigidities, and other factors would make it hard to replace less effective methods with the new method. Help from agency leaders, branch heads, and high-level administrators would be necessary to improve problem solving beyond individual efforts.

> "[We need] conscious attempts to apply process in the Branch in looking at problem solving. If it is consciously practised it is much more likely to become 2nd nature. A commitment by management to put issues through this framework will greatly facilitate gains made in the workshop."

> "This approach provides a framework within which most CNR activities can be reviewed. As such, it has potential use at most levels. Clearly, as the course has shown, it can usefully guide critical examination and evaluation of programs of enormous variation. Resourcing required in social science [skills], etc., to complement in-house expertise, time too is a resource scarcely allocated [by agency leaders] at present."

> "Reporting back on how our mid-year self-run session went could be useful [if we did one]. And it would help make sure we try the methods (including evaluation) before we come to the next course."

Long-Term Evaluation

The evaluations and recommendations were made by the participants at the conclusion of the workshops and within the following year and thus represent short-term appraisals. All feedback indicated that the workshops were highly successful, but long-term evaluations are needed. It is clear that the framework's analytic concepts and terms were used selectively. For example, the concepts of problem orientation and definition and social process were used by many attendees. It is also clear that the agency's working climate did not support either individual or group problem solving using this approach despite its wide endorsement by workshop participants.

From the standpoint of the instructor and managers, the workshops were successful. Attendees were well motivated, curious, and actively engaged with the new concepts and methods. They participated vigorously in discussions and case analyses. Because nearly all had considerable experience to contribute to the workshops, they were eager to present it constructively for the benefit of the group. In the workplace in the months following the workshops, managers and attendees reported that the workshop participants selectively used the concepts and methods and that the language used by the workers to describe their work and find solutions had changed.

It is now several years since the first workshop, and it is useful to ask about the outcomes and effects of the workshops in applications by participants. Future evaluations should determine long-term benefits to the participants: Were there any lasting benefits at the individual or agency levels? Have problem-solving methods improved? If so, how and with what effect? It is essential to determine if agency routines, cultures, and incentive systems have changed in response to the workshops. Is more workshop repetition required? Would more exposure to the concepts and opportunities for practical application be beneficial? Should additional workshops and official, intra-agency "reinforcing" exercises be held? If yes, what form should they take? By comparison with university-level courses of at least one hundred contact hours over four months with many cases and problems sets, how can a workshop of three or four days (twenty-four to thirty-two hours) be structured to impart knowledge and hone problem-solving skills?

Conclusions

The workshops clearly demonstrated that the interdisciplinary problem-solving method can be learned and applied relatively easily by agency professionals. This multi-year "ground testing" of workshops as a means to enhance professional problem solving was sufficiently successful to warrant future workshops in other professional communities. The workshops proved their practical benefits for the participants, including increased understanding of complex natural resource management problems and identification of ways to intervene constructively. The policy sciences problem-solving method can help users see the relevant context of the problems they face more reliably than they otherwise would using only conventional, discipline-based or organization-based standpoints. The method permits professionals to carry out a policy-oriented, functional analysis of any decision process, whether they are participants or observers. To master this conceptual framework in practical applications requires intelligence, integrity, hard work, practice, and persistence, characteristics that these workshop participants exhibited. In order to be most successful, future workshops should be carried out in agency environments that support the method and follow up with strong, long-term incentives to continue using it for better management outcomes.

Making connections between the general tools of interdisciplinary problem solving and particular conservation issues is difficult at first. Although the method is useful for any kind of problem, each problem is unique in its concrete details. So users must move back and forth between the general concepts and the case particulars. This requires professional judgment and interpretation—learning how to think critically and systematically—as well as practice in a supportive group setting. But the effort pays off as professionals perceive more of the problem-relevant context, gain confidence in their understanding of the details, and recognize patterns that had been overlooked.

ACKNOWLEDGMENTS

Workshops were conducted under a cooperative agreement between the Victoria Department of Natural Resources and Environment, Yale University, and the Northern Rockies Conservation Cooperative. Funding came from the Victoria Department of Natural Resources and Environment and from American sources via NRCC. The appendix manuscript was critically reviewed by Denise Casey and Andy Willard.

Robert Begg and Kim Lowe both work for the Department of Natural Resources and Environment in Melbourne, Victoria 3002, Australia. Begg is also a professor at Deakin University in Clayton, Victoria.

Literature Cited

Albee, G. W. 1982. Preventing psychopathology and promoting human potential. *American Psychologist* 37:1043–50.

Alcock, J. 1975. *Animal behavior: An evolutionary approach.* Sunderland, Mass.: Sinauer Associates.

Alvarez, K. 1993. *Twilight of the panther: Biology, bureaucracy and failure in an endangered species program.* Sarasota, Fla.: Myakka River Publishing.

———. 1994. The Florida panther recovery program: An organizational failure of the Endangered Species Act. Pp. 205–26 in *Endangered species recovery: Finding the lessons, improving the process,* edited by T. W. Clark, R. P. Reading, and A. L. Clarke. Washington: Island Press.

Argyris, C. 1992. *On organizational learning.* Cambridge, Mass.: Blackwell.

Argyris, C., and D. A. Schön. 1978. *Organizational learning: A theory of action perspective.* Reading, Mass.: Addison-Wesley.

Ascher, W. 1978. *Forecasting: An appraisal for policy-makers and planners.* Baltimore: Johns Hopkins University Press.

———. 1981. The forecasting potential of complex models. *Policy Sciences* 13: 247–67.

———. 1986. The evolution of the policy sciences: Understanding the rise and avoiding the fall. *Journal of Policy Analysis and Management* 5:367–73.

———. 1995. *Communities and sustainable forestry in developing countries.* San Francisco: Institute for Contemporary Studies.

Ascher, W., and R. Healy. 1990. *Natural resource policymaking in developing countries.* Durham, N.C.: Duke University Press.

Banff–Bow Valley Task Force. 1996. Banff–Bow Valley: At the crossroads. Summary Report of the Banff–Bow Valley Task Force to Minister of Canadian Heritage, Ottawa.

Barber, B. 1961. Resistance by scientists to scientific discovery. *Science* 134:596–602.

Bardach, E. 1977. *The implementation game: What happens after a bill becomes law.* Cambridge: MIT Press.

Barrett, W. 1978. *The illusion of technique: A search for meaning in a technological civilization.* Garden City, N.Y.: Anchor.

Barzun, J., and H. F. Graff. 1985. *The modern researchers.* New York: Harcourt Brace Jovanovich.

Batt, J., and D. Short. 1992–93. The jurisprudence of the 1992 Rio Declaration on environment and development: A law, science, and policy explication of certain aspects of the United Nations Conference on Environment and Development. *Journal of Natural Resources & Environment* 8:229–92.

Beardsley, E. L., and M. C. Beardsley. 1972. *Invitation to philosophical thinking.* New York: Harcourt Brace Jovanovich.

Bednarz, D., and D. J. Wood. 1991. *Research in teams: A practical guide to group policy analysis.* Englewood Cliffs, N.J.: Prentice-Hall.

Bem, D. J. 1970. *Beliefs, attitudes, and human affairs.* Belmont, Calif.: Brooks/Cole.

Bennett, A., G. N. Backhouse, and T. W. Clark, eds. *People and nature conservation: Perspectives on private land use and endangered species recovery.* Transactions of the Royal Zoological Society of New South Wales. Mosman, N.S.W.: Royal Zoological Society of New South Wales.

Bentley, J. W. 1994. Facts, fantasies, and failures of farmer participatory research. *Agriculture and Human Values* (spring–summer):140–50.

Berger, P. L. 1963. *Invitation to sociology: A humanistic perspective.* New York: Doubleday Anchor.

Berger, P. L., and T. Luckman. 1987. *Social construction of reality: A treatise in the sociology of knowledge.* New York: Penguin.

Berk, R. A., and P. H. Rossi. 1990. Key concepts in evaluation research. Pp. 12–32 in *Thinking about program evaluation.* London: Sage.

Berkes, F., and C. Folke, eds. 2000. *Linking social and ecological systems.* Cambridge: Cambridge University Press.

Berman, M. 1981. The reenchantment of the world. Ithaca: Cornell University Press.

———. 2000. *The twilight of American culture.* New York: Norton.

Berry, J. K. 2000. From paradigm to practice: Public involvement strategies for America's forests. Ph.D. diss., School of Forestry and Environmental Studies, Yale University.

Betts, R. K. 1978. Analysis, war, and decision: Why intelligence failures are inevitable. *World Politics* 31:61–89.

Beveridge, W. I. B. 1950. *The art of scientific investigation.* New York: Vintage.

Birkland, T. 1997. *After disaster: Agenda setting, public policy, and focusing events.* Washington, D.C.: Georgetown University Press.

Blanco, H. 1991. *How to think about social problems: American pragmatism and the idea of planning.* Westport, Conn.: Greenwood.

Bolland, J. M., and R. Muth. 1984. The decision seminar: A new approach to urban problem solving. *Knowledge: Creation, Diffusion, Utilization* 6:75–88.

Boyce, M. S., and E. M. Anderson. 1999. Evaluating the role of carnivores in the Greater Yellowstone Ecosystem. Pp. 265–84 in *Carnivores in ecosystems: The Yellowstone experience,* edited by T. W. Clark, A. P. Curlee, S. C. Minta, and P. M. Kareiva. New Haven: Yale University Press.

Brandon, K., K. H. Redford, and S. E. Sanderson. 1998. *Parks in peril: People, politics, and protected areas.* Washington, D.C.: Island.

Brewer, G. D. 1973a. Experimentation and the policy process. Pp. 151–163 in *25th Annual Report of the Rand Corporation.* Santa Monica, Calif.: Rand Corporation.

———. 1973b. The policy process as a perspective for understanding. Pp. 57–76 in *Children, families and government,* edited by E. Zigler, S. L. Kagan, and E. Klugman. New York: Cambridge University Press.

Brewer, G. D. 1973c. Professionalism: The need for standards. *Interfaces* 4:20–27.

———. 1974. The policy sciences emerge: To nurture and structure a discipline. *Policy Sciences* 5:239–44.

———. 1975. Dealing with complex social problems: The potential of the "decision seminar." Pp. 439–61 in *Political development and change: A policy approach,* edited by G. D. Brewer and R. D. Brunner. New York: Free.

———. 1978. Termination: Hard choices—harder questions. *Public Administration Review* (July–August):1–6.

———. 1981. Where the twain meet: Reconciling science and politics in analysis. *Policy Sciences* 13:269–79.

———. 1986. Methods for synthesis: Policy exercise. Pp. 455–75 in *Sustainable development of the biosphere,* edited by W. C. Clark and R. E. Munn. New Rochelle, N.Y.: Cambridge University Press.

———. 1995. Environmental challenges: Interdisciplinary opportunities and new ways of doing business. MISTRA Lecture, Royal Institute of Technology, Stockholm, Sweden.

Brewer, G. D., and T. W. Clark. 1994. A policy sciences perspective: Improving implementation. Pp. 391–413 in *Endangered species recovery: Finding the lessons, improving the process,* edited by T. W. Clark, R. P. Reading, and A. C. Clarke. Washington: Island.

Brewer, G. D., and P. deLeon. 1983. *The foundations of policy analysis.* Homewood, Ill.: Dorsey.

Brown, S. 1980. *Political subjectivity: Applications of Q methodology in political science.* New Haven: Yale University Press.

Bruner, J. 1990. *Acts of meaning.* Cambridge: Harvard University Press.

Brunner, R. D. 1982. The policy sciences as science. *Policy Sciences* 15:115–35.

———. 1991a. The policy movement as a policy problem. *Policy Sciences* 24:65–98.

———. 1991b. Global climate change: Defining the policy problem. *Policy Sciences* 24:291–311.

———. 1994. Myth and American politics. *Policy Sciences* 27:1–18.

———. 1995a. Notes on basic concepts of the policy sciences. Department of Political Science, University of Colorado, Boulder. Typescript.

———. 1995b. Harold D. Lasswell. Pp. 723–25 in *Encyclopedia of democracy,* edited by S. M. Lipset. Washington, D.C.: CQ.

———. 1996a. A milestone in the policy sciences. *Policy Sciences* 29:45–68.

———. 1996b. Policy and global change research: A modest proposal. *Climatic Change* 32:121–47.

———. 1996c. Policy sciences. *Social science encyclopedia.* 2d ed. London: Routledge.

———. 1997a. Raising standards: A prototyping strategy for undergraduate education. *Policy Sciences* 30:167–89.

———. 1997b. Teaching the policy sciences: Reflections on a graduate seminar. *Policy Sciences* 30:217–31.

———. 1997c. Barriers and bridges to the renewal of ecosystems and institutions: A review. *Journal of Wildlife Management* 61: 1437–39.

———. 1997d. Introduction to the policy sciences. *Policy Sciences* 30:191–215.

Brunner, R. D., and W. Ascher. 1992. Science and social responsibility. *Policy Sciences* 25:295–331.

Brunner, R. D., and T. W. Clark. 1997. A practice-based approach to ecosystem management. *Conservation Biology* 11:48–58.

Brunner, R. D., and R. Klein. 1999. Harvesting experience: A reappraisal of the U.S. Climate change action plan. *Policy Sciences* 32:133–61.

Burgess, P. M., and L. L. Slonaker. 1978. The decision seminar: A strategy for problem-solving. Merschon Center of the Ohio State University, Columbus. Paper No. 1.

Byers, B. A. 1996. Understanding and influencing behaviors in conservation and natural resources management. African Biodiversity Series, vol. 4. Washington, D.C.: Biodiversity Support Program.

Caputo, D. A. 1977. *The politics of policy making in America: Five case studies*. San Francisco: Freeman.

Carson, R. 1962. *Silent spring*. Boston: Houghton Mifflin.

Chen, L. 1989. *An introduction to contemporary international law: A policy-oriented perspective*. New Haven: Yale University Press.

Christensen, N. L., et al. 1996. The report of the Ecological Society of America Committee on the scientific basis for ecosystem management. *Ecological Applications* 6:665–93.

Clark, T. W. 1986a. Professional excellence in wildlife and natural resource organizations. *Natural Resources Journal* (summer):8–13.

———. 1986b. Case studies in wildlife policy education. *Renewable Resources Journal* 9 (4):11–16.

———. 1992. Practicing natural resource management with a policy orientation. *Environmental Management* 16:423–33.

———. 1993. Creating and using knowledge for species and ecosystem conservation: Science, organizations, and policy. *Perspectives in Biology and Medicine* 36: 497–525 and appendices.

———. 1996a. Appraising threatened species recovery efforts: Practical recommendations. Pp. 1–22 in *Back from the brink: Refining the threatened species recovery process*, edited by S. Stephens and S. Maxwell. Transactions of the Royal Zoological Society of New South Wales. Mosman, N.S.W.: Royal Zoological Society of New South Wales and Australian Nature Conservation Agency.

———. 1996b. Learning as a strategy for improving endangered species conservation. *Endangered Species Update* 13 (1–2):5–6,22–24.

———. 1997a. *Averting extinction: Reconstructing endangered species recovery*. New Haven: Yale University Press.

———. 1997b. Conservation biologists in the policy process: Learning how to be practical and effective. Pp. 575–97 in *Principles of conservation biology*, edited by G. K. Meffe and C. R. Carroll. 2nd ed. Sunderland, Mass.: Sinauer Associates.

———. 1999. Interdisciplinary problem-solving: Next steps in the Greater Yellowstone Ecosystem. *Policy Sciences* 32:393–414.

Clark, T. W., and M. Ashton. 1999. Field trips in natural resources professional education: The Panama case and recommendations. *Journal of Sustainable Forestry* 8:181–98.

Clark, T. W., G. N. Backhouse, and R. P. Reading. 1995. Prototyping in endangered species recovery programmes: The eastern barred bandicoot experience. Pp. 50–62 in *People and nature conservation: Perspectives on private land use and endangered species recovery*, edited by A. Bennett, G. N. Backhouse, and T. W. Clark. Transactions of the Royal Zoological Society of New South Wales. Mosman, N.S.W.: Royal Zoological Society of New South Wales.

Clark, T. W., and G. D. Brewer. 2000. Introduction. Pp. 9–22 in *Developing sustainable management policy for the National Elk Refuge, Wyoming*, edited by T. W. Clark, D. Casey, and A. Halverson. Yale School of Forestry and Environmental Studies Bulletin no. 104.

Clark, T. W., and R. D. Brunner. 1996. Making partnerships work in endangered species conservation: An introduction to the decision process. *Endangered Species Update* 13 (9):1–5.

Clark, T. W., D. Casey, and A. Halverson, eds. 2000. *Developing sustainable management policy for the National Elk Refuge, Wyoming.* Yale School of Forestry and Environmental Studies Bulletin no. 104.

Clark, T. W., R. Crete, and J. Cada. 1989. Designing and managing successful endangered species recovery programs. *Environmental Management* 13:159–70.

Clark, T. W., A. P. Curlee, and R. P. Reading. 1996. Crafting effective solutions to the large carnivore conservation problem. *Conservation Biology* 10:940–48.

Clark, T. W., A. H. Harvey, M. B. Rutherford, B. Suttle, S. Primm, and A. P. Curlee. 1999. *Annotated bibliography on management of the Greater Yellowstone Ecosystem.* Jackson, Wyo.: Northern Rockies Conservation Cooperative.

Clark, T. W., N. Mazur, R. J. Begg, and S. J. Cork. 2000. Interdisciplinary guidelines for developing effective koala conservation policy. *Conservation Biology* 14 (3): 691–701.

Clark, T. W., N. Mazur, S. Cork, S. Dovers, and R. Harding. 2000. The koala conservation policy process: An appraisal and recommendations. *Conservation Biology* 14 (3):681–90.

Clark, T. W., and S. C. Minta. 1994. *Greater Yellowstone's future: Prospects for ecosystem science, management, and policy.* Moose, Wyo.: Homestead.

Clark, T. W., and R. P. Reading. 1994. A professional perspective: Improving problem solving, communication, and effectiveness. Pp. 351–70 in *Endangered species recovery: Finding the lessons, improving the process,* edited by T. W. Clark, R. P. Reading, and A. C. Clarke. Washington: Island.

Clark, T. W., and R. L. Wallace. 1999. The professional in endangered species conservation: An introduction to standpoint clarification. *Endangered Species Update* 16 (1):9–13.

Clark, T. W., and A. R. Willard. 2000. Analyzing natural resources policy and management. Pp. 32–44 in *Foundations of natural resources policy and management,* edited by T. W. Clark, A. R. Willard, and C. M. Cromley. New Haven: Yale University Press.

Clark, T. W., A. R. Willard, and C. M. Cromley, eds. 2000. *Foundations of natural resources policy and management.* New Haven: Yale University Press.

Clark, W. C. 1986. Sustainable development of the biosphere: Themes for a research program. Pp. 5–48 in *Sustainable development of the biosphere,* edited by W. C. Clark and R. E. Munn. New Rochelle, N.Y.: Cambridge University Press.

Clark, W. C., et al. 1990. *Managing planet earth: Readings from* Scientific American *Magazine.* New York: Freeman.

Clarke, J. N., and D. McCool. 1985. *Staking out the terrain: Power differentials among natural resource management agencies.* Albany: State University of New York Press.

Cole, C. F. 1992. New dimensions to the education of fish and wildlife professionals. *Environmental Professional* 14:325–32.

Cork, S.J., T. W. Clark, and N. Mazur. 2000. Special section: Conservation of koalas in Australia. *Conservation Biology* 14: 606–704.

Cortner, H. J., and M. A. Moote. 1998. *The politics of ecosystem management.* Washington: Island.

Coughlan, B. A. K., and C. L. Armour. 1992. Group decision-making techniques for natural resource management applications. U.S. Department of the Interior, Fish and Wildlife Service, Resource Publication 185. Washington, D.C.: U.S. Goverment Printing Office.

Court of International Trade. 1995. Earth Island Institute v. Cristopher, 913 F. Supp. 559, CIT, New York.

Cromley, C. M. 2002. Beyond boundaries: Learning from bison management in Greater Yellowstone. Ph.D. diss., School of Forestry and Environmental Studies, Yale University.

Crosthwaite, J. 1995. Identifying and addressing social and economic issues in recovery planning: Experience with the Victorian Flora and Fauna Guarantee Act 1988. Pp. 78–86 in *People and nature conservation: Perspectives on private land use and endangered species recovery,* edited by A. Bennett, G. N. Backhouse, and T. W. Clark. Transactions of the Royal Zoological Society of New South Wales. Mosman, N.S.W.: Royal Zoological Society of New South Wales.

Culhane, R. J. 1981. *Public lands politics: Resources for the future.* Baltimore: Johns Hopkins University Press.

Dahlberg, K. 1983. Contextual analysis: Taking space, time, and place seriously. *International Studies Quarterly* 27:257–66.

Davis, G., J. Wanna, J. Warhurst, and P. Weller. 1988. *Public policy in Australia.* North Sydney: Allen and Unwin.

Denzin, N. K., and Y. S. Lincoln, eds. 1994. *Handbook of qualitative research.* Thousand Oaks, Calif.: Sage.

Dery, D. 1984a. *Problem definition in policy analysis.* Lawrence: University of Kansas Press.

———. 1984b. Evaluation and termination in the policy cycle. *Policy Sciences* 17: 13–26.

Dession, G. H., and H. D. Lasswell. 1955. Public order under law: The role of the advisor-draftsman in the formation of code or constitution. *Yale Law Journal* 65:174–95.

Dewey. J. 1981–90. *The later works, 1925–1953,* edited by J. A. Boydston. 17 volumes. Carbondale: Southern Illinois University Press.

Dey, I. 1993. *Qualitative data analysis: A user-friendly guide for social scientists.* New York: Routledge.

Dillard, A. 1982. The problem of knowing the world. Pp. 53–56 in *Living by fiction.* New York: Harper and Row.

Dobyns, H. F., P. L. Doughty, and H. D. Lasswell, eds. 1971. *Peasants, power, and applied social change.* Beverly Hills: Sage.

Dominowski, R. L. 1980. *Research methods.* Englewood Cliffs, N.J.: Prentice-Hall.

Doob, L. W. 1995. *Sustainers and sustainability: Attitudes, attributes, and actions for survival.* Westport, Conn.: Praeger.

Doughty, P. L. 1987. Against the odds: Collaboration and development at Vicos. Pp. 129–57 in *Collaboration research and social change: Applied anthropology in action,* edited by D. D. Stull and J. J. Schensul. Boulder, Colo.: Westview.

Dovers, S. R. 1996. Policy process for sustainability. Ph.D. diss., Australian National University, Canberra.

Dror, E. 1971a. Series editor's introductory note. P. xi in H. D. Lasswell, *A preview of policy sciences.* New York: American Elsevier.

———. 1971b. *Ventures in policy sciences: Concepts and applications.* Ventura, Calif.: American Elsevier.

Dryzek, J. S. 1990. *Discursive democracy: Politics, policy, and political science.* Cambridge: Cambridge University Press.

Dunn, W. N. 1981. *Policy analysis: Perspectives, concepts, and methods.* Greenwich, Conn: JAI.

Durning, D. 1996. The transition from traditional to postpositivistic policy analysis: A role for Q methodology. Eighteenth Annual Research Conference of the Association of Public Policy Analysis and Management, Pittsburgh.

Edge, A. 1991. *The guide to case analysis and reporting.* Honolulu: Systems Logistics.

Ehrlich, P. R. 1997. *World of wounds: Ecologists and the human dilemma.* Oldendorf-Luhe, Germany: Ecology Institute.

Etheredge, L. 1985. *Can governments learn?* New York: Pergamon.

Fesler, J. W. 1980. Implementation: Success and failure. Pp. 248–77 in *Public administration: Theory and practice.* Englewood Cliffs, N.J.: Prentice-Hall.

Fish, S. 1996. Professor Sokal's bad joke. *New York Times,* 21 May:A13.

Flannery, T. F. 1994. *The future eaters: An ecological history of Australasian lands and people.* Chatswood, New South Wales: Reed.

Fleck, L. 1979. *Genesis and development of a scientific fact.* Chicago: University of Chicago Press.

Flores, A. 2000. Protecting human health from ozone pollution in Baltimore, Maryland: Revising the current policy. Pp. 47–79 in *Foundations of natural resources policy and management,* edited by T. W. Clark, A. R. Willard, and C. M. Cromley. New Haven: Yale University Press.

Foundation for Strategic Environmental Research and The Council for Planning and Co-ordination of Research. 1998. The theory and practice of interdisciplinary work. Conference proceedings, June 8–10, Stockholm.

Freudenberger, K. S. 1997. Rapid rural appraisal and participatory rural appraisal. Notes to accompany an introductory course. School of Forestry and Environmental Studies, Yale University.

Friedman, M. 1953. The methodology of positive economics. Pp. 3–43 in *Essays in positive economics.* Chicago: University of Chicago Press.

Friedmann, J., and H. Rangan. 1993. *In defense of livelihood: Comparative studies on environmental action.* West Hartford, Conn.: Kumarian.

Funtowicz, S. O., and J. R. Ravetz. 1990. *Uncertainty and quality in science for policy.* Dordrecht: Kluwer.

Garen, E. 2000. Appraising ecotourism in conserving biodiversity. Pp. 221–51 in *Foundations of natural resources policy and management,* edited by T. W. Clark, A. R. Willard, and C. M. Cromley. New Haven: Yale University Press.

Gilovich, T. 1991. *How we know what isn't so: The fallibility of human reasoning in everyday life.* New York: Free.

Goleman, D. 1995. *Emotional intelligence: Why it can matter more than IQ.* New York: Bantam.

Golembiewski, R. T., and M. White. 1983. *Cases in public management.* Boston: Houghton Mifflin.

Goldman, A. I. 1986. *Epistemology and cognition.* Cambridge: Harvard University Press.

Grumbine, E. 1990. Protecting biological diversity through the greater ecosystem concept. *Natural Areas Journal* 10:114–10.

———. 1994. What is ecosystem management? *Conservation Biology* 8 (1):27–38.

Gunderson, L. H., C. S. Holling, and S. S. Light, eds. 1995. *Barriers and bridges to the renewal of ecosystems and institutions.* New York: Columbia University Press.

Habermas, J. 1987. *The philosophical discourse of modernity.* Boston: Beacon.

Haeuber, R., and J. Franklin. 1996. Perspectives on ecosystem management. *Ecological Applications* 6:692–93.

Hall, P. 1993. Policy paradigms, social learning, and the state: The case of economic policymaking in Britain. *Comparative Politics* 25 (3):275–96.

Halpern, D. F. 1996. *Thinking critically about critical thinking.* Mahwah, N.J.: Lawrence Erlbaum Associates.

Ham, C., and M. Hill. 1986. *The policy process in the modern capitalistic state.* Sussex, U.K.: Wheatsheaf.

Hanna, S. 1994. Summary of a workshop on property rights and natural resources. *Society and Natural Resources* 7:595–97.

Harre, R. 1985. *The philosophies of science.* Oxford: Oxford University Press.

Healy, R. G., and W. Ascher. 1995. Knowledge in the policy process: Incorporating new environmental information in natural resource policy making. *Policy Sciences* 28:1–19.

Heberlein, T. A. 1988. Improving interdisciplinary research: Integrating the social and natural sciences. *Society and Natural Resources* 1:5–16.

Hein, D. 1995. Traditional education in natural resources. Pp. 75–87 in *A new century for natural resources management,* edited by R. L. Knight and S. F. Bates. Washington, D.C.: Island.

Hogwood, B. W., and L. A. Gunn. 1987. *Policy analysis for the real world.* Oxford: Oxford University Press.

Holling, C. S. 1995. What barriers? What bridges? Pp. 3–36 in *Barriers and bridges to the renewal of ecosystems and institutions,* edited by L. H. Gunderson, C. S. Holling, and S. S. Light. New York: Columbia University Press.

Horgan, J. 1996. *The end of science: Facing the limits of knowledge in the twilight of the scientific age.* New York: Broadway.

International Institute for Environment and Development. 1994. Whose Eden? An overview of community approaches to wildlife management. A report to the Overseas Development Administration of the British Government, London.

International Union for the Conservation of Nature and Natural Resources, United Nations Environment Program, and World Wildlife Fund. 1991. *Caring for the earth: A strategy for sustainable living.* Gland, Switz.: IUCN.

International Union for the Conservation of Nature and Natural Resources, World Wildlife Fund, and United Nations Environment Program. 1980. *World conservation strategy: Living resource conservation for sustainable development.* Gland, Switz.: IUCN.

Isaac, S., and W. B. Michael. 1995. *Handbook in research and evaluation: A collection of principles, methods, and strategies useful in the planning, design, and evaluation of studies in education and behavioral sciences.* San Francisco: Educational and Industrial Testing Services.

James, W. 1978. *Pragmatism and the meaning of truth.* Cambridge: Harvard University Press.

Jamieson, D. 1991. The epistemology of climate change: Some morals for managers. *Society and Natural Resources* 4:319–29.

Janesick, V. J. 1994. The dance of qualitative research design: Metaphor, methodology, and meaning. Pp. 209–19 in *Handbook of qualitative research,* edited by N. K. Denzin and Y. S. Lincoln. Thousand Oaks, Calif.: Sage.

Janis, I. 1972. *Victims of groupthink.* Boston: Houghton Mifflin.

Johnson, D. M. 1965. *The international law of fisheries: A framework for policy-oriented inquires.* New Haven: Yale University Press.

Johnson, R. H. 1992. The problem of defining critical thinking. Pp. 38–53 in *The generalizability of critical thinking,* edited by S. P. Norris. New York: Teachers College Press.

Kates, R. W., B. L. Turner II, and W. C. Clark. 1990. The great transformation. Pp. 1–18 in *The earth as transformed by human action: Global and regional changes in the biosphere over the past 300 years.* New York: Cambridge University Press.

Katz, K., and R. L. Kahn. 1966. *The social psychology of organizations*. New York: Wiley.

Kuhn, T. S. 1962. *The structure of scientific revolutions*. Chicago: Phoenix.

―――. 1977. *The essential tension: Selected studies in scientific tradition and change*. Chicago: University of Chicago Press.

Lasch, C. 1978. *The culture of narcissism: American life in an age of diminishing expectations*. New York: Norton.

―――. 1991. *The true and only heaven: Progress and its critics*. New York: Norton.

Lasswell, H. D. 1938. Intensive and extensive methods of observing the personality-culture manifold. *Yenching Journal of Social Studies* 1:72–86.

―――. 1950a. *Politics: Who gets what, when, how*. New York: Peter Smith.

―――. 1950b. *World politics and personal insecurity*. Glencoe, Ill.: Free.

―――. 1951a. Democratic character. Pp. 465–525 in *The political writings of Harold D. Lasswell*. Glencoe, Ill.: Free.

―――. 1951b. The policy orientation. Pp. 3–15 in *The policy sciences: Recent developments in scope and method*, edited by D. Lerner and H. D. Lasswell. Stanford: Stanford University Press.

―――. 1956. The decision process: Seven categories of functional analysis. College Park, Md.: Bureau of Governmental Research and College of Business and Public Administration, University of Maryland.

―――. 1959. Strategies of inquiry: The rational use of observation. Pp. 89–113 in *The human meaning of the social sciences*, edited by D. Lerner. New York: Meridian-World.

―――. 1960a. *Psychopathology and politics*. New York: Viking.

―――. 1960b. Technique of decision seminar. *Midwest Journal of Political Science* 4:213–36.

―――. 1963. *The future of political science*. New York: Atherton.

―――. 1966a. *The analysis of political behavior: An empirical approach*. Hamden, Conn.: Archon.

―――. 1966b. Decision seminars: The contextual use of audiovisual means in teaching, research, and consultation. Pp. 499–524 in *Comparing nations: The use of quantitative data in cross-national research*, edited by R. L. Merritt and S. Rokkan. New Haven: Yale University Press.

―――. 1968. Policy sciences. *International Encyclopedia of the Social Sciences* 12: 181–89.

―――. 1970. The emerging conception of the policy sciences. *Policy Sciences* 1: 3–14.

―――. 1971a. *A pre-view of policy sciences*. New York: American Elsevier.

―――. 1971b. *The policy orientation of political sciences*. Agra, India: Lakshmi Narain Agarwal.

―――. 1971c. From fragmentation to configuration. *Policy Sciences* 29:45–68.

―――. 1971d. The continuing decision seminar as a technique of instruction. *Policy Sciences* 2:43–57.

―――. 1971e. Sharing the experience of permanent reconstruction: A policy science approach. Pp. 536–46 in *Essays on modernization of underdeveloped societies*, edited by A. R. Desai. Bombay: Thacker.

―――. 1994. Introduction: Universality versus parochialism. Pp. lxxxiii–lxxxix in *The international law of war: Transnational coercion and world public order*, edited by M. S. McDougal and F. P. Feliciano. New Haven: New Haven Press.

Lasswell, H. D., and M. B. Fox. 1979. *The signature of power: Buildings, communication, and policy*. New Brunswick, N.J.: Transaction.

Lasswell, H. D., and A. R. Holmberg. 1992. Toward a general theory of directed value accumulation and institutional development. Pp. 1379–1417 in *Jurisprudence for a free society: Studies in law, science, and politics*, edited by H. D. Lasswell and M. S. McDougal. New Haven: New Haven Press.

Lasswell, H. D., and A. Kaplan. 1950. *Power and society*. New Haven: New Haven Press.

Lasswell, H. D., and M. S. McDougal. 1992. *Jurisprudence for a free society: Studies in law, science, and policy*. 2 vols. New Haven: New Haven Press.

Latour, B. 1987. *Science in action*. Cambridge: Harvard University Press.

Lee, K. N. 1993. *The compass and gyroscope: Integrating science and politics for the environment*. Washington: Island.

Lerner, D., and H. D. Lasswell, eds. 1951. *The policy sciences*. Stanford: Stanford University Press.

Lerner, M. 1996. *The politics of meaning: Restoring hope and possibility in an age of cynicism*. Reading, Mass.: Addison-Wesley.

Lewin, A. Y., and M. F. Shakun. 1976. *Policy sciences: Methodologies and cases*. New York: Pergamon.

Lewontin, R. C. 1992. *Biology as ideology*. New York: HarperPerennial.

Lichtman, P., and T. W. Clark 1994. Rethinking the "Vision" exercise in the Greater Yellowstone Ecosystem. *Society and Natural Resources* 7:459–78.

Lindblom, C. E. 1959. The science of muddling through. *Public Administration Review* 19:79–88.

———. 1980. *The policy-making process*. Englewood Cliffs, N.J.: Prentice-Hall.

Lindblom, C. E., and E. J. Woodhouse. 1993. *The policy-making process*. 3d ed. Englewood Cliffs, N.J.: Prentice-Hall.

Lippman, W. 1965. *Public opinion*. New York: Free.

Little, T. M., and F. J. Hills. 1978. *Agricultural experimentation: Design and analysis*. New York: John Wiley and Sons.

Lubchenco, J. 1998. Entering the century of the environment: A new social contract for science. *Science* 279:491–97.

Lynn, L. E., Jr. 1980. *Designing public policy: A casebook on the role of policy analysis*. Santa Monica: Goodyear.

Machlis, G. E., J. E. Force, and W. R. Burch Jr. 1997. The human ecosystem. Part I: The human ecosystem as an organizing concept in ecosystem management. Cooperative Park Studies Unit, University of Idaho, Moscow.

MacIver, R. M. 1947. *The web of government*. New York: Macmillan.

Maguire, L. A. 1986. Using decision analysis to manage endangered species populations. *Journal of Environmental Management* 22:345–60.

———. 1995. Desired future conditions for the Chattanooga Watershed: A qualitative analysis of diverse public opinions. Report to the Southeast Forest Experiment Station, U.S. Forest Service.

Maier, N. R. F. 1962. *Problem-solving discussion and conferences: Leadership methods and skills*. New York: McGraw-Hill.

Majchrzak, A. 1984. *Methods for policy research*. Applied Social Research, vol. 3. Beverly Hills: Russell Sage.

Majone, G. 1989. *Evidence, argument, and persuasion in the policy process*. New Haven: Yale University Press.

Mander, J. 1991. *In the absence of the sacred: The failure of technology and the survival of the Indian nations*. San Francisco: Sierra Club Books.

Manheim, H. L. 1977. *Sociological research: Philosophy and methods*. Homewood, Ill.: Dorsey.

Mansergh, I., A. Jelinek, and P. Clunie. 1995. A review of the action statement process under the Victorian Flora and Fauna Guarantee Act 1988. Pp. 68–77 in *People and nature conservation: Perspectives on private land use and endangered species recovery,* edited by A. Bennett, G. N. Backhouse, and T. W. Clark. Transactions of the Royal Zoological Society of New South Wales. Mosman, N.S.W.: Royal Zoological Society of New South Wales.

Manwell, C., and C. M. A. Baker. 1979. The double helix: Science and myth in the act of creation. *BioScience* 29:742–56.

March, J. G., and H. A. Simon. 1961. *Organizations.* New York: John Wiley and Sons.

Marcus, G. E., and M. F. Fischer. 1988. *Anthropology as cultural critique: An experimental moment in the human sciences.* Chicago: University of Chicago Press.

Mares, M. A. 1991. How scientists can impede the development of their discipline: Ego-centrism, small pool size, and evolution of sapismo. Pp. 57–75 in *Latin American mammalogy: History, biodiversity, and conservation,* edited by M. Mares and D. J. Schmidley. Norman: Oklahoma Museum of Natural History.

Margoluis, R., and N. Salafsky. 1998. *Measures of success: Designing, managing, and monitoring conservation and development projects.* Washington: Island.

Marius, R. 1995. *A short guide to writing about history.* New York: HarperCollins.

Maslow, A. 1970. *Motivation and personality.* 2d ed. New York: Van Nostrand Reinhold.

———. 1971. *The farther reaches of human nature.* New York: Viking.

May, R. 1991. *The cry for myth.* New York: Norton.

McCain, G., and E. M. Segal. 1977. *The game of science.* Monterey, Calif: Brooks/ Cole.

McDougal, M. S. 1992a. Human rights and world public order: Principles of content and procedure for clarifying general community policies. Pp. 1527–64 in *Jurisprudence for a free society: Studies in law, science, and policy,* edited by H. D. Lasswell and M. S. McDougal. New Haven: New Haven Press.

———. 1992b. McDougal on Lasswell. Pp. xxix–xxxiv in *Jurisprudence for a free society: Studies in law, science and policy,* edited by H. D. Lasswell and M. McDougal. New Haven Press, New Haven.

———. 1992–93. Legal basis for securing the integrity of the earth-space environment. *Journal of Natural Resources and Environmental Law* 8:177–207.

McDougal, M. S., H. D. Lasswell, and L. Chen. 1980. *Human rights and world public order: Basic policies of an international law of human dignity.* New Haven: Yale University Press.

McDougal, M. S., H. D. Lasswell, and W. M. Reisman. 1981. Theories about international law: Prologue to a configurative jurisprudence. Pp. 43–141 in *International law essays: A supplement to international law in contemporary perspective,* edited by M. S. McDougal and W. M. Reisman. Mineola, N.Y.: Foundation.

McDougal, M. S., and W. M. Reisman. 1981. Constitutive process. Pp. 269–86 in *International law essays: A supplement to international law in contemporary perspective,* edited by M. S. McDougal and W. M. Reisman. Mineola, N.Y.: Foundation.

McDougal, M. S., W. M. Reisman, and A. R. Willard. 1988. The world community: A planetary social process. *University of California Law Review* 21:807–972.

McKeown, B., and D. Thomas. 1988. *Q methodology.* Methods Series, vol. 3. Beverly Hills: Russell Sage.

Meltsner, A. J. 1976. *Policy analysis in the bureaucracy.* Berkeley: University of California Press.

Michael, D. N. 1995. Barriers and bridges to learning in a turbulent human ecology. Pp. 461–88 in *Barriers and bridges to the renewal of ecosystems and institu-*

tions, edited by L. H. Gunderson, C. S. Holling, and S. S. Light. New York: Columbia University Press.

Miewald, R., and S. Welch. 1983. Natural resources: An introduction. Pp. 9–24 in *Scarce natural resources: The challenge to public policymaking,* edited by S. Welch and R. Miewald. Beverly Hills: Russell Sage.

Miller, A. 1985. Cognitive styles and environmental problem-solving. *International Journal of Environmental Studies* 26:21–31.

———. 1999. *Environmental problem solving: Psychosocial barriers to adaptive management.* New York: Springer.

Miller, B. J., R. P. Reading, and S. C. Forrest. 1996. *In the dying light: Recovery of the black-footed ferret.* Washington, D.C.: Smithsonian Institution Press.

Miller, D. C. 1991. *Handbook of research design and social measurement.* Thousand Oaks, Calif.: Sage.

Miller, K. R. 1996. *Balancing the scales: Guidelines for increasing biodiversity's chances through bioregional management.* Washington, D.C.: World Resources Institute.

Miser, R. 1956. *Positivism: A study of human understanding.* New York: George Braziller.

Molnar, A. 1989. *Community forestry: Rapid appraisal.* Rome: Food and Agriculture Organization of the United Nations.

Moore, J. N. 1968. Prolegomenon to the jurisprudence of Myres McDougal and Harold Lasswell. *Virginia Law Review* 54:662–88.

Morgan, G. 1986. *Images of organization.* Beverly Hills: Sage.

Muth, R. 1987. The decision seminar: A problem-solving technique for school administrators. *Planning and Changing* 18:45–60.

Muth, R., and J. M. Bolland. 1981. The social planetarium: Toward a revitalized civic order. *Urban Interest* 3 (2):13–25.

Muth R., M. M. Finley, and M. F. Muth. 1990. *Harold D. Lasswell: An annotated bibliography.* New Haven: New Haven Press.

Nakamura, R. T., and F. Smallwood. 1980. *The politics of policy implementation.* New York: St. Martin's.

National Research Council. 1986. *Ecological knowledge for environmental problem solving.* Washington: National Academy Press.

Norse, E. A., ed. 1993. *Global marine biological diversity: A strategy for building conservation into decision making.* Washington: Island.

Noss, R. F., M. A. O'Connell, and D. D. Murphy. 1997. *The science of conservation planning: Habitat conservation under the Endangered Species Act.* Washington: Island.

Olsen, M. E., D. G. Lodwick, and R. E. Dunlap. 1992. *Viewing the world ecologically.* Boulder, Colo.: Westview.

Paehlke, R., ed. 1995. Sustainability, sustainable agriculture, sustainable development, sustained yield forestry. Pp. 612–18 in *Conservation and environmentalism: An encyclopedia.* New York: Garland.

Parson, E. A., and W. C. Clark. 1995. Sustainable development as social learning: Theoretical perspectives and practical challenges for the design of a research program. Pp. 428–60 in *Barriers and bridges to the renewal of ecosystems and institutions,* edited by L. H. Gunderson, C. S. Holling, and S. S. Light. New York: Columbia University Press.

Parsons, W. 1995. *Public policy: An introduction to the theory and practice of policy analysis.* Lyme, Conn.: Edward Elgar.

Patai, R. 1972. *Myth and modern man.* Englewood Cliffs, N.J.: Prentice-Hall.

Perrow, C. 1984. *Normal accidents: Living with high-risk technologies*. New York: Basic.

Pickett, S. T. A., R. S. Ostfeld, M. Harchak, and G. E. Lickens. 1997. *The ecological basis of conservation: Heterogeneity, ecosystems, and biodiversity*. New York: Chapman and Hall.

Pimbert, M. P., and J. N. Pretty. 1995. Parks, people, and professionals: Putting "participation" into protected area management. United Nations Research Institute for Social Development, Discussion Paper DP 57.

Popper, M., and R. Lipshitz. 1998. Organizational learning mechanisms: A structural and cultural approach to organizational learning. *Journal of Applied Behavioral Science* 34:161–79.

Pressman, J. L., and A. Wildvasky. 1979. *Implementation*. Berkeley: University of California Press.

Primm, S. A. 2000. Real bears, symbol bears, and problem solving. *Northern Rockies Conservation Cooperative News* (Jackson, Wyoming), 13:6–8.

Primm, S. A, and T. W. Clark. 1996. The Greater Yellowstone policy debate: What is the policy problem? *Policy Sciences* 29:137–66.

Putt, A. D., and J. F. Springer. 1989. *Policy research: Concepts, methods, and applications*. Englewood Cliffs, N.J.: Prentice-Hall.

Pye-Smith, J., G. B. Feyerband, and R. Sandbrook. 1994. *The wealth of communities: Stories of success in local environmental management*. West Hartford, Conn.: Kumarian.

Quade, E. S. 1975. *Analysis for public decisions*. New York: Elsevier.

Raffensperger, C., and J. Tickner. 1999. *Protecting public health and the environment*. Washington: Island.

Rappaport, R. A. 1979. *Ecology, meaning, and religion*. Richmond, Calif.: North Atlantic.

Reisman, W. M. 1969. *Development and social change: An overview*. Yale Law School, Yale University. Typescript.

———. 1981a. Law from the policy perspective. Pp. 1–14 in *International law essays: A supplement to international law in contemporary perspective*, edited by M. S. McDougal and W. M. Reisman. Mineola, N.Y.: Foundation.

———. 1981b. International lawmaking: A process of communication. *World Academy of Art and Science* 1981:101–20.

———. 1981c. Private armies in a global war system: Prologue to decision. Pp. 142–90 in *International law essays: A supplement to international law in contemporary perspective*, edited by M. S. McDougal and W. M. Reisman. Mineola, N.Y.: Foundation.

Reisman, W. M., and A. R. Willard. 1988. *International incidents: The law that counts in world politics*. Princeton: Princeton University Press.

Richards, S. 1983. *Philosophy and sociology of science: An introduction*. Padstow, U.K.: Basil Blackwell.

Robar, S. F. 1999. Collaboration, policy perspectives and discursive democracy: Public land management and the Colorado Plateau Forum. Ph.D. diss., Department of Political Science, Northern Arizona University, Flagstaff.

Rochefort, D. A., and R. W. Cobb. 1994. *The politics of problem definition*. Lawrence: University of Kansas Press.

Rosaldo, R. 1993. *Culture and truth: The remaking of social analysis*. Boston: Beacon Press.

Rose, H., and S. Rose, eds. 1976. *Ideology in the natural sciences*. Cambridge, Mass.: Schenkman.

Rose, R. 1993. *Lesson drawing in public policy: A guide to learning across time and space.* Chatham, N.J.: Chatham House.

Rossi, P. H., and H. E. Freeman. 1993. *Evaluation.* 5th edition. Thousand Oaks, Calif.: Sage Publications.

Sabatier, P., and H. Jenkins-Smith. 1999. The advocacy coalition framework: An assessment. Pp. 26–44 in *Theories of policy process,* edited by P. Sabatier. Boulder, Colo.: Westview.

Sachs, A. 1995. Eco-justice: Linking human rights and the environment. Worldwatch Paper 127. Washington, D.C.: Worldwatch Institute.

Sarewitz, D., and R. Pielke, Jr. 2000. Breaking the global-warming gridlock. *Atlantic Monthly,* July:55–64.

Schneider, A. L., and H. Ingram. 1997. *Policy design for democracy.* Lawrence: University of Kansas Press.

Schön, D. A. 1983. *The reflective practitioner: How professionals think in action.* New York: Basic.

Scott, J. C. 1998. *Seeing like a state: How certain schemes to improve the human condition have failed.* New Haven: Yale University Press.

Sellars, R. W. 1997. *Preserving nature in the national parks: A history.* New Haven: Yale University Press.

Senge, P. M. 1990. *The fifth discipline: The art and practice of the learning organization.* New York: Doubleday.

Shrader-Frechette, K. S., and E. D. McCoy. 1994. Ecology and environmental problem-solving. *Environmental Professional* 16: 342–48.

Shutkin, W. 1999. *The land that could be: Environmentalism and democracy in the twenty-first century.* Cambridge: MIT Press.

Simon, H. A. 1983. *Reason in human affairs.* Stanford: Stanford University Press.

———. 1985. Human nature in politics: The dialogue of psychology with political science. *American Political Science Review* 79:293–304.

———. 1996. *The sciences of the artificial.* Cambridge: MIT Press.

Singer, F. J., and J. A. Mack. 1999. Predicting the effects of wildfire and carnivore predation on ungulates. Pp. 189–238 in *Carnivores in ecosystems: The Yellowstone experience,* edited by T. W. Clark, A. P. Curlee, S. C. Minta, and P. M. Kareiva. New Haven: Yale University Press.

Slotkin, R. 1992. *Gunfighter nation: The myth of the frontier in twentieth-century America.* New York: Harper Perennial.

Smith, K. K., and D. N. Berg. 1987. *Paradoxes of group life: Understanding conflict, paralysis, and movement in group dynamics.* San Francisco: Jossey-Bass.

Sokal, A. 1996. A physicist experiments with cultural studies. *Lingua Franca* (May–June):62–64.

Somluckrat, W. G., T. B. Grandstaff, and G. W. Lovelace. 1985. Summary report. Pp. 1–30 in *Proceedings of the 1985 International Conference on Rapid Rural Appraisal.* Khon Kaen University, Thailand.

Starling, G. 1988. *Strategies for policy making.* Chicago: Dorsey.

Steelman, T. A. 1997. Q-methodology: A tool for facilitating public involvement in natural resource management. Sixteenth Annual Policy Sciences Institute, Yale University Law School, New Haven.

Stephenson, W. 1964. Application of Q-method to the measurement of public opinion. *Psychological Record* 14:265–73.

Strauss, A., and J. Corbin. 1994. Grounded theory. Pp. 273–85 in *Handbook of qualitative research,* edited by N. K. Denzin and Y. S. Lincoln. Thousand Oaks, Calif.: Sage.

Sullivan, W. M. 1995. *Work and integrity: The crises and promise of professionalism in America*. New York: Harper Business.

Taylor, C. 1989. *Sources of the self: The making of the modern identity*. Cambridge: Harvard University Press.

Taylor, J. L. 1983. Developing environmental management from a case-study base. *Environmental Conservation* 9:261–63.

Thant, U. 1969. Report of the Secretary-General, problems of the human environment. United Nations Economic and Social Council, May 26, 1969, E/4667.

Thomas, J. W. 1986. Effectiveness—the hallmark of the natural resource management professional. *Transactions of the North American Wildlife and Natural Resources Conference* 51:27–38.

Thomas, W. L., Jr. 1956. *Man's role in changing the face of the earth*. Chicago: University of Chicago Press.

Thompson, G. C. 1966. The evaluation of public opinion. Pp. 7–12 in *Reader in public opinion and communication*, edited by B. Berelson and M. Janowitz. 2d ed. New York: Free.

Torgerson, D. 1985. Contextual orientation in policy analysis: The contribution of Harold D. Lasswell. *Policy Sciences* 18:241–61.

Tribe, L. 1973. Policy sciences: Analysis or ideology? *Philosophy and Public Affairs* 2:66–110.

Tufte, E. R. 1990. *Envisioning information*. Cheshire, Conn.: Graphics.

———. 1992. *The visual display of quantitative information*. Cheshire, Conn.: Graphics.

———. 1997. *Visual explanations: Images and quantities, evidence and narrative*. Cheshire, Conn.: Graphics.

Turner, F. 1980. *Beyond geography: The western spirit against the wilderness*. New York: Viking.

United Nations Environmental Programme. 1992a. *Saving our planet: The state of the environment (1972–1992)*. Nairobi: UNEP.

———. 1992b. *Agenda 21: The United Nations programme of action from Rio*. New York: United Nations.

United States Department of the Interior. 1993. New director foresees ecosystem management, new partnerships in agency's future. News release, FWS, November 3, 1993. INT REQ 4–00548.

United States Fish and Wildlife Service. 1994. An ecosystem approach to fish and wildlife conservation: An approach to more effectively conserve the nation's biodiversity. Washington, D.C.: U.S. Fish and Wildlife Service.

Vickers, G. 1965. *The art of judgment: A study of policy making*. London: Chapman and Hall.

Warren, M. E. 1999. *Democracy and trust*. New York: Cambridge University Press.

Weimer, D. L., and A. R. Vining. 1989. *Policy analysis: Concepts and practice*. Englewood Cliffs, N.J.: Prentice-Hall.

Weiss, J. A. 1989. The powers of problem definition: The case of government paperwork. *Policy Sciences* 22:92–121.

Westrum, R. 1986. Management strategies and information failure. NATO Advanced Research Workshop on Failure Analysis of Information Systems, August, Bad Winsheim, Germany.

———. 1994. An organizational perspective: Designing recovery teams from the inside out. Pp. 327–49 in *Endangered species recovery: Finding the lessons, improving the process*, edited by T. W. Clark, R. P. Reading, and A. L. Clarke. Washington: Island.

White, A. T., L. Z. Hale, Y. Renard, and L. Cortesi, eds. 1994. *Collaborative and community-based management of coral reefs: Lessons from experience*. West Hartford, Conn.: Kumarian.

White, L. 1967. The historical roots of our ecological crises. *Science* 155:1203–7.

Wiener, P. P., ed. 1958. *Charles S. Peirce, Selected writing: Values in a universe of chance*. New York: Dover.

Wildvasky, A. 1979. *Speaking truth to power: The art and craft of policy analysis*. Boston: Little, Brown.

Willard, A. R., and C. H. Norchi. 1993. The decision seminar as an instrument of power and enlightenment. *Political Psychology* 14:575–606.

Williams, G. W. 1994. Ecosystem management: How did we get here? Paper presented at Society of American Foresters, Indianapolis, Indiana, February 22:1–16.

Williams, R. M., Jr. 1979. Change and stability in values and value systems: A sociological perspective. Pp. 15–46 in *Understanding human values: Individual and societal*, edited by M. Rokeach. New York: Free.

Wilson, B., and T. W. Clark. 1995. The Victorian Flora and Fauna Guarantee Act 1988: A five year review of its implementation. Pp. 50–62 in *People and nature conservation: Perspectives on private land use and endangered species recovery*, edited by A. Bennett, G. N. Backhouse, and T. W. Clark. Transactions of the Royal Zoological Society of New South Wales. Mosman, N.S.W.: Royal Zoological Society of New South Wales.

Winks, R. W. 1997. The National Park Service Act of 1916: "A contradictory mandate?" *Denver University Law Review* 74:575–623.

Wondolleck, J. M., and S. L. Yaffee. 2000. *Making collaboration work: Lessons from innovation in natural resource management*. Washington: Island.

World Commission on Environment and Development. 1987. *Our common future*. New York: Oxford University Press.

World Conservation Monitoring Centre. 1992. *Global biodiversity: Status of the Earth's living resources*. London: Chapman and Hall.

World Resources Institute. 1994. *World resources 1994–95: A guide to the global environment*. New York: Oxford University Press.

World Resources Institute, World Conservation Union, and United Nations Environment Programme. 1992. *Global biodiversity strategy*. Washington: World Resources Institute.

Yaffee, S. L. 1982. *Prohibitive policy: Implementing the federal Endangered Species Act*. Cambridge: MIT Press.

———. 1994a. *The wisdom of the spotted owl: Policy lessons for a new century*. Washington: Island Press.

———. 1994b. The northern spotted owl: An indicator of the importance of sociopolitical context. Pp. 47–71 in *Endangered species recovery: Finding the lessons, improving the process*, edited by T. W. Clark, R. P. Reading, and A. L. Clarke. Washington: Island Press.

Yin, R. K. 1989. *Case study research: Design and methods*. Applied Social Research Methods, vol. 5. London: Sage.

Zerner, C. 2000. *People, plants and justice: The politics of nature conservation*. New York: Columbia University Press.

Index

Action Statements, application function and, 68

adaptive management, principle of contextuality and, 29–30

adaptive perspective, to evaluate alternatives, 98–99

advocacy, as a policy exercise, 148

affection, defined, 27

Agenda 21: Programme of Action for Sustainable Development, 63

Albee, George, 1

alternatives, in problem orientation, 96–100; information gathering in context of, 104

analytic framework. *See* framework *and* policy sciences framework

application, as a decision function, 66–68

appraisal: as a decision function, 68–69, 115–17; rapid rural appraisal, 150–51; participatory rural appraisal, 150–51

Argyris, Chris, 166, 169

Ascher, William, 5–6, 22, 71, 93, 95, 121–22, 133

assets, in prescriptions, 64

"At the Crossroads," 109

authority signal, in prescription, 64

Babbit, Bruce, 173

Baker, C. M. A., 121, 171

Banff–Bow Valley Round Table, 108–9

Banff–Bow Valley Task Force case study, 107–9

base values: in logging case study, 50–51; methods for mapping, 45–46; in social process, 34, 40–41; in world social process mapping, 53–54

Batt, John, 28, 37

Begg, Robert J., interdisciplinary workshop of, in Australia, 77, 173–89

behavior: analyzing conditions and, 92–94; contextuality of, 29–30; of groups and societies, 19–21; of individuals, 17–19; maximization postulate and, 24–25; perspectives and, 35; social science methods to investigate, 47

best practices, in resource management, 144–45

bias, policy-oriented professionalism and, 112–13

biopolitics, compared to policy, 6

bison management, 3

bounded rationality, 24–25

Brewer, G. D., 3, 87, 117, 128–29, 174; on conventional disciplines, 8; on decision process, 59; on policy exercises, 145–48

Brundtland Commission. *See* World Commission on Environment and Development (WCED)